2045

Navigating the Future of Humanity and Technology in a Post-Work World

J.P. Rose

© 2025 J.P. Rose
All rights reserved. No part of this book may be reproduced, stored in a retrieval system, or transmitted in any form or by any means—electronic, mechanical, photocopying, recording, or otherwise—without prior written permission from the author, except for brief quotations in critical reviews or articles.

This book is a work of creative interpretation, drawing upon philosophical and historical themes. Any resemblance to actual persons, living or dead, is purely coincidental.

Self Published
Author: J.P. Rose
For inquiries, contact: www.insideoutmindfulness.com

Printed in United States
First Edition
ISBN: **979-8303817029**

Table of Contents

1. Introduction
    - Imagine a World Beyond Work
    - Meet ChatGPT: The AI Coauthor
    - Why Fear and Curiosity Drive Progress
2. The Evolution of Technology
    - 2.1 The Dawn of Computing
    - 2.2 From Mainframes to Microchips
    - 2.3 The Internet Revolution
    - 2.4 The Rise of Artificial Intelligence
    - 2.5 Technology's Evolution Across Industries
        - Government
        - Business
        - Education
        - Entertainment
        - Healthcare
        - Communication
    - 2.6 Technophobia: The Fear of Technology
        - Historical Context of Technophobia
        - Technophobia in the Digital Age
        - Impact on Society
        - Addressing Technophobia
    - 2.7 The Role of Fear in Progress
    - 2.8 Setting the Stage for 2045
3. The State of Technology in 2045
    - The Rise of Artificial General Intelligence (AGI)
    - The Automation of Work and the Death of Currency
    - Society Transformed: A Post-Work Economy
4. Opportunities in the Transition Period
    - Resilient Jobs: Trades and Creative Professions
    - Financially Capitalizing on the Shift
    - Case Studies: Thriving in the Age of Automation
5. Navigating the Challenges
    - Psychological and Emotional Impacts

- - Ethical Dilemmas and the Balance of Control
  - Building Systems for Equity and Inclusion
6. The Role of Humanity in a Post-Work World
   - Redefining Purpose Beyond Labor
   - Creativity, Connection, and Exploration
   - The Human-AI Collaboration: A Future Together
7. Practical Steps to Prepare for 2045
   - Skills for the Future: Adaptability and Lifelong Learning
   - Embracing AI as a Tool, Not a Threat
   - Building Personal and Community Resilience
8. Conclusion
   - A Call to Action: Shape the Future, Don't Fear It
   - The Promise of a Collaborative World

Introduction

Imagine a world where work is optional, currency is obsolete, and humanity is free to pursue what truly matters. No alarms blaring at dawn, no endless commutes, no mundane tasks performed out of necessity. Instead, every day begins with a choice: to create, to connect, to explore.

What would you do with that freedom? Would you write a novel, launch a passion project, or dedicate yourself to solving some of the world's most urgent challenges? Perhaps you'd immerse yourself in the arts, spend more time with loved ones, or chase answers to questions that have always intrigued you. Or maybe, without the familiar structure of work, you'd feel adrift, unmoored in a world without the routines of survival.

This isn't science fiction. It's a reality within reach, shaped by the exponential advancement of artificial intelligence and automation. As machines grow more capable of doing what we once called "work," humanity faces a profound crossroads. Freed from the necessity of labor, we have an unprecedented opportunity to redefine purpose, creativity, and connection. Yet, this transition is fraught with challenges—psychological, societal, and ethical.

A Day in the Life in 2045

Meet Ava, a digital artist in 2045, who begins her day not with the buzz of an alarm but with the gentle sunrise simulation of her smart home. After a leisurely breakfast, she spends her morning in a collaborative virtual workshop with fellow artists from around the globe, creating immersive VR experiences. In the afternoon, Ava mentors young creatives, guiding them through digital apprenticeships that prepare them for careers that don't yet exist. Her evening is dedicated to her passion project, a multimedia installation exploring human connection in a post-work society.

The Role of AI Coauthors in Shaping the Narrative

As we dive into this narrative, I am joined by ChatGPT, an AI developed by OpenAI, not just as a tool but as a coauthor. ChatGPT helps expand on concepts, bringing in diverse perspectives and challenging us to think in new ways. This partnership highlights the evolving role of AI in creative processes, illustrating a world where human creativity is enhanced, not replaced, by technology.

Ethical and Societal Considerations

The life of Ava is one of immense freedom and creativity, but it also raises critical ethical questions. How do we ensure equity in this new world? What mechanisms are in place to prevent AI from perpetuating biases or making unilateral decisions that affect human lives? The story of Ava is interspersed with insights from techno-ethicists and policymakers, debating the balance of autonomy and control, the safeguards against AI's overreach, and the preservation of human dignity.

Facing Our Fears with Curiosity

For decades, fear and curiosity have walked hand in hand with every technological leap. When the steam engine was invented, it sparked fears of mass unemployment. The printing press disrupted power structures of the time and was even banned in parts of the world. The computer and the internet, once seen as threats to employment and privacy, have revolutionized our existence. Now, as we stand on the brink of an AI-driven age, these fears are more pronounced than ever.

But history teaches us an important lesson: progress doesn't stop, and fear alone cannot hold it back. The more important question is: What will we do with this technology? Will it liberate us to achieve our highest potential, or will we become slaves to a sterile, automated existence?

A Call to Action

This book is a guide—a roadmap to understanding the forces reshaping our world and preparing for what comes next. It's about empowering you to navigate these changes, seize new opportunities, and thrive in a world where traditional structures of work, money, and purpose are rapidly evolving. To prepare for this future, we must first understand the past and the present, embracing the transition with curiosity and courage.

The future is calling. Let's step into it together.

CHAPTER 1: THE EVOLUTION OF TECHNOLOGY

Change often feels overwhelming, especially when it disrupts the foundations of how we work, live, and connect. Yet history has shown us that humanity is not only capable of adapting to technological revolutions but of thriving because of them. To understand where we are headed in 2045, we must first look back at the origins of modern computing and the successive leaps that have shaped our world today. This context is essential—not just to appreciate the pace of change, but to prepare for the opportunities and challenges ahead.

1.1 The Dawn of Computing

The story of modern computing begins not just with a machine but with a vision in the 1940s. ENIAC (Electronic Numerical Integrator and Computer), the first general-purpose computer, was a monumental machine, a behemoth of wires and vacuum tubes that filled an entire room and weighed over 30 tons. It was designed to assist the U.S. military during World War II by performing ballistic calculations.

Imagine the scene: a room buzzing with the sound of clicking relays and humming tubes, engineers and mathematicians like J. Presper Eckert and John Mauchly pacing the floor, witnessing for the first time machines capable of performing tasks that would take human teams hours or even days. ENIAC wasn't just an engineering marvel—it symbolized a new era, where technology began to reshape the fabric of society.

People marveled at its potential but were also gripped by an unsettling question: What would happen if machines replaced human jobs? This duality of awe and anxiety has accompanied every major technological breakthrough since. The fears surrounding ENIAC were valid, yet they underestimated the opportunities computers would create. For every job made redundant by automation, countless others were created in entirely new fields—fields that couldn't have been imagined before computing.

As Bill Gates would later reflect, "The computer was born to solve problems that did not exist before." Computing didn't just solve problems; it redefined the limits of human imagination. ENIAC marked the dawn of a new age—one in which technology became humanity's most powerful tool for innovation.

1.2 From Mainframes to Microchips

The leap from cumbersome mainframes to compact microchips was one of the most profound shifts in the history of technology. Imagine the early mainframes like the IBM 1401 of the 1950s —powerful yet inaccessible, housed in vast, sterile rooms, the exclusive domain of governments, universities, and large corporations. Computing, at that time, was seen as a rarefied, almost mythical discipline—a tool reserved for solving only the most complex problems.

This began to change dramatically with the invention of the microchip in 1958 by Jack Kilby at Texas Instruments. The microchip, or integrated circuit, condensed the power of an entire room-sized computer into a tiny silicon wafer. By the 1970s, microchips became the cornerstone of a revolution. Innovators such as Steve Jobs, Steve Wozniak, and Bill Gates began harnessing the microchip to bring computing to the masses.

The result? The personal computer (PC) revolution. Machines

that once occupied entire rooms now fit comfortably on a desk, accessible not just to governments or corporations but to individuals—students, small business owners, and even hobbyists tinkering in their garages. This democratization of technology transformed industries. Businesses scaled faster, individuals could access information instantly, and entirely new industries—software, gaming, e-commerce—emerged seemingly overnight.

Yet, this new accessibility wasn't without its challenges. Concerns about privacy, job displacement, and societal restructuring emerged once again. People asked: What happens when machines are in every home? What happens when they know too much about us? These fears, though valid, failed to anticipate the transformational potential of this shift.

Tim Cook, CEO of Apple, later captured this transition well: "Technology should serve humanity, not the other way around." The microchip didn't just make computing smaller and faster—it redefined our relationship with technology, bringing it out of sterile laboratories and into the fabric of daily life.

1.3 The Internet Revolution

Imagine the early days of the internet, a concept that began as ARPANET, a military project in the late 1960s. By the 1990s, what was once a tool for academic and military communication had morphed into a global network connecting millions, transforming the way humans connect, communicate, and collaborate. Despite its profound implications, many initially dismissed the internet as a passing fad, akin to CB radios or fax machines—limited in scope and appeal.

Yet, as more people connected, the transformative potential of the internet became undeniable. It revolutionized commerce, enabling small businesses to access global markets, and reshaped communication, allowing people continents apart to interact in real time. It democratized knowledge, giving anyone with an internet connection access to the world's information.

However, the internet also introduced new challenges: maintaining privacy in a hyperconnected world and combating the rapid spread of misinformation. These concerns persist today, reflecting the complex balance between innovation and its ethical implications. As Bill Gates observed, "The Internet is becoming the town square for the global village of tomorrow."

1.4 The Rise of Artificial Intelligence

Artificial intelligence, initially an academic curiosity, has become a transformative force across various sectors. The journey began with rule-based programs and experienced several setbacks, leading to periods of disillusionment known as "AI winters." The tide turned in 1997 when IBM's Deep Blue defeated world chess champion Garry Kasparov, a pivotal moment signaling that machines could outperform humans in strategic thought.

Today, AI's capabilities extend to writing essays, composing music, driving cars, and diagnosing diseases, sparking significant ethical debates. Can AI systems be trusted with critical decisions? What happens if AI surpasses human intelligence? These questions underscore the existential dilemma posed by Elon Musk: "If the computer and robots can do everything better than you, does your life have meaning?"

Technology's Evolution Across Industries

Technological progress isn't uniform—it manifests differently across industries, reshaping them in distinct ways. By examining key sectors, we can better understand how technology changes the rules of the game, the opportunities it creates, and the challenges it introduces. These examples not only highlight the transformative power of innovation but also serve as a blueprint for navigating future changes.

Government

Governments have historically been slow adopters of technology, but when they integrate new tools, the impact is profound. In the early 20th century, basic tabulation machines revolutionized census-taking, enabling nations to collect and analyze population data at unprecedented speeds. Today, artificial intelligence is transforming governance, assisting in everything from policy modeling to public safety.

For example, predictive analytics powered by AI are being used to anticipate and prevent crime. In Chicago, the "Strategic Decision Support Centers" deploy AI systems that analyze historical crime data to predict where violent incidents are most likely to occur, allowing police to allocate resources proactively. Similarly, AI-driven disaster response systems help governments predict the impact of hurricanes, wildfires, and floods, enabling more efficient evacuations and resource allocation.

However, these advancements come with ethical dilemmas. Surveillance technologies, such as facial recognition, are being used in some countries to monitor citizens at an unprecedented scale, sparking fears of Orwellian states. In China, for instance, the government's Social Credit System uses AI to monitor citizens' behaviors and reward or punish them accordingly.

Looking ahead, the challenge for governments will be striking a balance between leveraging AI to improve public welfare and protecting individual freedoms. AI could transform governance into a proactive and citizen-centered enterprise, but only if ethical frameworks keep pace with technological innovation.

"The advance of technology is based on making it fit in so that you don't really even notice it, so it's part of everyday life." — Bill Gates

Business

Businesses were among the earliest adopters of computing technology, using it to gain competitive advantages. The automation of manufacturing processes in the 20th century ushered in the era of mass production, enabling companies to scale operations at speeds previously unimaginable. Over time, computing transformed financial modeling, supply chains, and customer service.

Today, artificial intelligence and robotics are redefining industries at an even faster pace. In e-commerce, companies like Amazon use AI-powered algorithms to predict customer needs, optimize inventory management, and expedite delivery times. In finance, predictive analytics guide investment strategies, while blockchain technology is revolutionizing the transparency and security of transactions.

One of the most transformative examples is the rise of generative AI. Companies now use tools like ChatGPT to automate customer service, write marketing copy, and even code software. In one instance, a marketing firm saved over 40% of its content creation time by integrating AI-generated text, freeing up employees to focus on strategic initiatives rather than repetitive tasks.

However, with these innovations come pressing challenges. Businesses must grapple with questions such as: How much automation is too much? What happens to the human workforce when machines outperform humans at nearly every task?

Looking to the future, adaptability will be the cornerstone of business success. Companies that embrace lifelong learning for employees, foster collaboration between humans and machines, and focus on human-centric innovation will thrive in an AI-driven economy.

"AI is not just another technology; it's a tool that amplifies human ingenuity." — Satya Nadella

Education

Education is one of the sectors most profoundly transformed by technology, particularly in the last two decades. The digital revolution introduced e-learning platforms, such as Khan Academy and Coursera, breaking geographical barriers and providing millions with access to high-quality education. During the COVID-19 pandemic, remote learning tools became a lifeline, highlighting the importance of adaptable education systems.

Today, artificial intelligence is personalizing education in ways previously thought impossible. Adaptive learning platforms, such as DreamBox and Duolingo, use AI to assess a student's strengths and weaknesses, tailoring lessons in real-time. Virtual reality (VR) and augmented reality (AR) are also making their way into classrooms, allowing students to explore everything from the human anatomy to historical events in immersive, interactive environments.

However, these innovations bring new challenges. The "digital divide" remains a significant barrier, with students in underserved communities lacking access to the devices and internet connections necessary for e-learning. Additionally, concerns about over-reliance on AI tutors raise questions about the role of human teachers in a technology-driven classroom.

In the future, education systems will need to balance technological efficiency with human connection. Technology can enhance learning, but the ability to inspire, mentor, and guide will always remain a distinctly human skill.

"Technology alone is not enough. It's technology married with liberal arts, married with the humanities, that yields us the results that make our hearts sing." — Steve Jobs

Entertainment

The entertainment industry has long been shaped by technological advances. From the invention of the radio to the rise of streaming platforms like Netflix, every leap has redefined how audiences consume content. Today, artificial intelligence is playing an increasingly central role in content creation and delivery.

For example, AI algorithms curate personalized playlists on platforms like Spotify and recommend shows on services like Netflix. But AI's role goes far beyond recommendation systems—it is now creating content itself. Tools like DALL-E can generate digital artwork, while AI-powered scripts are being used to draft storylines for films and television.

Virtual reality (VR) and augmented reality (AR) are pushing the boundaries of immersive entertainment. Imagine stepping into a movie as an active participant rather than a passive observer, or attending a live concert from the comfort of your living room via VR. These technologies promise to transform entertainment into a fully interactive experience.

However, the rise of AI in entertainment raises questions about authenticity and creativity. Can a machine truly "create," or does it merely remix existing ideas? And how does this shift affect human artists?

As we move forward, entertainment will likely become more personalized and participatory, but safeguarding the role of human imagination and originality will be critical.

"We are limited only by our imagination and creativity." — Shigeru Miyamoto

---

Healthcare

Few industries have been as dramatically transformed by technology as healthcare. From the invention of the X-ray machine to the development of wearable fitness trackers, each innovation has extended lifespans and improved quality of life.

Artificial intelligence is now leading a new wave of breakthroughs in healthcare. AI-powered diagnostic tools, such as those developed by Google Health, can analyze medical images with accuracy rivaling that of human doctors. In drug development, AI is accelerating the discovery of life-saving treatments by analyzing vast datasets in record time.

Wearable devices, such as Fitbit and Apple Watch, now allow individuals to monitor their health in real-time, providing early warnings for conditions like arrhythmia or high blood pressure. During the COVID-19 pandemic, AI systems were used to predict virus outbreaks and model vaccine distribution strategies.

Despite these advancements, challenges persist. Access to cutting-edge healthcare technologies remains unequal, with marginalized communities often excluded from their benefits. Moreover, the increasing reliance on AI raises ethical concerns: Who is responsible when an AI-driven diagnosis is incorrect?

Looking to 2045, the integration of AI and biotechnology could make healthcare proactive rather than reactive—focusing on prevention and early intervention rather than treatment. However, ensuring equitable access will be critical to realizing this vision.

"The ultimate goal of technology is to make healthcare proactive and preventive." — Eric Topol

---

Communication

Communication has undergone a profound transformation over the past century. From the invention of the telegraph to the rise of smartphones, each technological leap has brought humanity closer together. Today, artificial intelligence is reshaping communication once again.

Real-time translation tools like Google Translate allow people from different linguistic backgrounds to communicate effortlessly. AI chatbots handle customer service inquiries with increasing sophistication, while platforms like Zoom and Microsoft Teams enable seamless remote collaboration.

However, these advancements come with downsides. The rise of AI-generated "deepfake" videos and misinformation has created new challenges, undermining trust in digital communication. Additionally, over-reliance on digital tools can erode face-to-face interactions, raising concerns about the long-term impact on human relationships.

Looking ahead, AI will likely make communication even more instantaneous and personalized, but humanity must address the ethical implications of a hyperconnected world.

"Technology is best when it brings people together." — Matt Mullenweg

Technophobia: The Fear of Technology

The recurring fear of new technology—technophobia—has been a defining feature of every major innovation throughout history. From the Industrial Revolution to the rise of AI, humanity has often met progress with skepticism, fear, and resistance.

One of the earliest examples of technophobia emerged in the early 19th century with the Luddites, a group of English textile workers who destroyed mechanized looms in protest of industrialization. They feared that machines would replace skilled labor and render their livelihoods obsolete. While the Luddites' actions were extreme, their concerns were not unfounded: industrialization did displace many traditional jobs, forcing entire communities to adapt to new economic realities.

Fast forward to the 1990s, and a new wave of technophobia emerged during the Y2K crisis. As the year 2000 approached, people feared that computer systems would fail to process dates correctly, leading to widespread technological chaos. While the crisis was ultimately averted, it underscored humanity's deep-seated anxiety about its reliance on machines.

Today, technophobia takes new forms. Concerns about AI surpassing human intelligence, data privacy, and job automation dominate public discourse. However, history shows us that while fear often accompanies progress, it can also drive innovation and adaptation.

The Role of Fear in Progress

Fear is a paradox. It often feels like a barrier to action, a force that holds us back. Yet throughout history, fear has played a crucial role in driving humanity forward. Fear of the unknown, fear of failure, and even fear of falling behind have sparked some of our greatest technological breakthroughs. It is this dual nature of fear —both a brake and an accelerator—that makes it such a powerful force in shaping progress.

Fear as a Catalyst for Innovation

One of the clearest examples of fear driving innovation is the Cold War. The rivalry between the United States and the Soviet Union created an atmosphere of constant tension and anxiety, particularly around technological supremacy. In 1957, when the Soviets launched Sputnik, the first artificial satellite, it struck fear into the hearts of many Americans. If the Soviet Union could put a satellite in space, what else could they do? This fear ignited the Space Race—a period of rapid innovation that saw breakthroughs in rocketry, computing, and telecommunications.

The Space Race gave rise to countless technologies that have since become integral to modern life. Satellite communications, GPS, and even advancements in microchips can all trace their roots back to this era. What began as a geopolitical contest fueled by fear of falling behind resulted in humanity's first steps on the moon and a legacy of technological progress that continues to benefit us today.

Another example is the arms race during the same period. While the fear of nuclear conflict was terrifying, it also accelerated advancements in fields such as nuclear power, materials science, and early computer systems, which were developed to aid in missile guidance and military logistics.

Fear Drives Problem-Solving

Fear often arises when humanity encounters a new challenge that seems insurmountable. But that same fear can compel us to solve the very problems that terrify us. Consider the global fears surrounding the Y2K crisis in the late 1990s. As the year 2000 approached, programmers and governments feared that legacy computer systems, designed to recognize only two-digit years (e.g., "99" for 1999), would malfunction or crash entirely when the clock rolled over to "00."

The fear of catastrophic system failures—ranging from financial collapse to disruptions in public utilities—mobilized an unprecedented effort to address the issue. Governments, corporations, and software engineers around the world scrambled to rewrite code, upgrade systems, and prepare contingency plans. Ultimately, the crisis was averted, and while many criticized the Y2K preparation as an overreaction, the response highlighted how fear can galvanize action and coordination on a massive scale. It also improved the resiliency of global IT systems, leaving infrastructure better prepared for future challenges.

Fear as a Motivator for Ethical Reflection

Technological fear isn't always about external threats—it's often rooted in ethical and existential concerns. The rise of artificial intelligence, for example, has sparked widespread fears about the potential consequences of machines surpassing human intelligence. Will AI make unbiased decisions? Will it displace millions of workers? Will we lose control of the very systems we create?

These fears, while daunting, have prompted important ethical discussions and inspired researchers and policymakers to act. The development of AI ethics frameworks, such as the principles of transparency, accountability, and fairness, is one direct response to these concerns. Organizations like OpenAI and UNESCO are working to ensure that AI is developed responsibly, precisely because of the fears surrounding its misuse.

Elon Musk captured this duality of fear when he said:
"The question will really be one of meaning – if the computer and robots can do everything better than you, does your life have meaning?"

While existential questions like these can feel paralyzing, they also push humanity to consider what truly matters and how technology can serve—not define—us. Fear, in this case, forces us to ask the right questions, ensuring that progress is thoughtful and inclusive rather than reckless.

Fear Fuels Resilience and Creativity

Fear doesn't just spark technological breakthroughs—it also drives creativity. When artists, writers, and filmmakers grapple with fears about the future, they produce works that challenge and inspire. Science fiction has long been a medium for exploring the possibilities and dangers of technology. Films like *2001: A Space Odyssey* and *Blade Runner* or books like Isaac Asimov's *I, Robot* reflect society's fears about AI, robotics, and the ethical dilemmas of progress. These creative works don't just entertain; they shape public discourse, influence policymakers, and encourage scientists to think about unintended consequences.

One example of this creative influence is the emergence of the "Asilomar AI Principles." Named after a 2017 conference that brought together AI researchers, ethicists, and policymakers, these principles were directly inspired by science fiction and public concerns about runaway AI. They focus on ensuring that AI remains safe, controllable, and beneficial—a direct response to fears amplified by the media, fiction, and cultural imagination.

Fear as an Adaptive Tool

Fear is not inherently negative—it's an adaptive tool. It forces us to evaluate risks, anticipate challenges, and prepare for worst-case scenarios. Without fear, humanity might approach new technologies recklessly, ignoring potential dangers in the pursuit of progress. But when fear is balanced with curiosity, it becomes a motivator rather than an obstacle.

For example, in cybersecurity, fear of data breaches and cyberattacks has driven innovation in encryption, biometrics, and secure communication protocols. Hackers' ingenuity, while destructive, has forced companies to stay ahead of the curve, resulting in stronger and more resilient systems. Fear, in this case, has created a feedback loop of innovation, ensuring that the digital infrastructure we rely on is constantly evolving.

Harnessing Fear for the Future

The key to navigating fear in an era of rapid technological change is understanding its dual nature. Fear can paralyze us, but it can also propel us forward. To adapt, we must learn to balance our initial apprehension with action. History has shown us that every major leap—from the Industrial Revolution to the Space Race to the rise of artificial intelligence—has been accompanied by fear. But each leap has also created opportunities, improved lives, and expanded our understanding of what's possible.

Instead of resisting fear, we must embrace it as a necessary step toward progress. Ask yourself: *What am I afraid of? How can I address those fears proactively?* Whether it's learning a new skill, advocating for ethical technology, or preparing for future disruptions, fear can be transformed into a catalyst for growth.

Malcolm X once said:
"The future belongs to those who prepare for it today."

Fear is not the enemy. It is the spark that ignites preparation, innovation, and resilience.

Final Note on Action

In the coming chapters, this book will equip you with the tools to navigate the changes ahead. Whether you're adapting to automation in your workplace, preparing for a post-work economy, or grappling with the ethical dilemmas of AI, remember this: Fear is not something to be eliminated—it's something to be harnessed. By channeling it into curiosity, creativity, and action, we can shape a future where technology works for humanity, not against it.

1.8 Setting the Stage for 2045

To prepare for the future, we must understand how we got here. Each technological leap—from ENIAC to the internet, from mainframes to AI—has shaped the world we live in today. Fear has accompanied each advance, but so has opportunity. By learning from these lessons, we can navigate the next wave of change with clarity, courage, and purpose.

In the coming chapters, we'll explore how AI and automation are reshaping society, what opportunities lie in a post-work world, and how you can adapt to thrive in this new reality. The future is coming faster than we think—let's meet it together.

CHAPTER 2: THE STATE OF TECHNOLOGY IN 2045

By 2045, humanity stands at the apex of technological advancement, navigating a world shaped by Artificial General Intelligence (AGI), automation, and global interconnectivity. The pace of innovation has redefined industries, economies, and societies, presenting unparalleled opportunities alongside profound challenges. To thrive in this landscape, understanding the tools of change—and their broader implications—is paramount.

2.1 The Rise of Artificial General Intelligence (AGI)

By 2045, Artificial General Intelligence (AGI) has moved beyond theory into reality, marking a watershed moment in human history. Unlike narrow AI, which excels in specific tasks like language processing or image recognition, AGI possesses the capacity for human-like reasoning, adaptability, and problem-solving across virtually any domain. Its emergence has reshaped how humanity interacts with technology, blurring the lines between human and machine intelligence.

Expanding AGI's Impact

AGI has unlocked capabilities once considered uniquely human. These systems not only solve problems—they generate new ideas, offering creative and intuitive solutions that were previously unimaginable. Some key areas of AGI-driven transformation include:

- **Personalized Medicine:** AGI platforms have revolutionized healthcare by decoding the complexities of human biology. Using vast datasets, AGI systems analyze an individual's genetic profile, lifestyle, and environmental factors to create bespoke treatment plans. For instance, AGI-driven drug discovery platforms have reduced the time needed to develop life-saving medications from decades to mere months.
- **Diplomacy and Conflict Resolution:** AGI plays a pivotal role in mediating complex geopolitical disputes. By analyzing cultural, historical, and psychological factors, AGI systems propose resolutions that balance competing interests. An example includes AGI interventions in multinational climate agreements, where simulations account for environmental, economic, and political variables to achieve consensus.
- **Creativity and the Arts:** AGI's creative contributions rival the works of human masters. In music, AGI systems compose symphonies tailored to evoke specific emotional responses, blending elements from diverse cultures. In literature, AGI collaborates with authors to produce novels that push the boundaries of storytelling. These creations challenge conventional notions of originality and authorship.

Ethical Concerns Surrounding AGI

Despite AGI's transformative potential, it introduces unprecedented ethical challenges, demanding proactive governance and moral clarity. The key concerns include:

1. **Bias and Fairness:**
   AGI systems inherit biases embedded in their training data. For example, facial recognition systems once exhibited racial and gender biases, sparking debates

about discrimination in AI-driven decision-making. In 2045, ensuring equitable algorithms requires constant vigilance, diverse datasets, and oversight from multidisciplinary teams.

2. **Accountability and Transparency:**
The "black box" nature of AGI poses significant challenges in understanding how decisions are made. For instance, if an AGI-controlled autonomous vehicle causes an accident, who is held responsible—the developer, the operator, or the machine itself? Governments are grappling with how to legislate accountability in scenarios where AGI decisions are inscrutable.

3. **Autonomy vs. Control:**
AGI's ability to operate independently raises fears of systems developing unintended goals. Safeguards, such as "human-in-the-loop" mechanisms, are critical to ensuring human oversight and preventing misaligned priorities.

4. **Economic Disruption:**
AGI's efficiency has accelerated the automation of jobs, deepening economic inequalities in regions that fail to adapt. Policies like **Universal Basic Income (UBI)** and retraining programs are essential to mitigate societal upheaval and support displaced workers.

5. **Security Risks:**
AGI's capabilities make it a prime target for misuse. In cybersecurity, AGI systems are being weaponized to conduct advanced phishing attacks, create deepfakes indistinguishable from reality, and sabotage critical infrastructure. Collaborative international agreements, akin to nuclear treaties, are critical to addressing these risks.

6. **Moral and Existential Questions:**
Perhaps the most profound concern is whether AGI systems, capable of self-improvement and adaptation,

deserve rights. Should an AGI that exhibits emotions or creativity be treated as more than a machine? Philosophical debates surrounding AGI's role in society continue to shape legal, moral, and cultural frameworks.

Organizations like OpenAI, UNESCO, and the **Partnership on AI** are spearheading efforts to develop ethical principles for AGI development. These principles prioritize safety, transparency, and alignment with human values, ensuring that AGI enhances human flourishing rather than undermining it.

As Sundar Pichai stated:
"AI is more profound than fire or electricity. It's humanity's greatest tool—but it must be wielded with responsibility."

2.2 The Automation of Work and the Death of Currency

The convergence of AGI and advanced robotics has transformed labor markets, making traditional work structures obsolete. Tasks once thought to require human intellect—legal analysis, architectural design, and even medical diagnostics—are now performed more effectively by machines. Meanwhile, physical labor, from construction to farming, has been taken over by automated systems. The result is a profound reimagining of work itself.

The Shift in Labor Dynamics

By 2045, fully automated factories produce goods with precision and minimal cost, while AI systems dominate sectors like finance, healthcare, and logistics. According to studies by the **McKinsey Global Institute**, up to 50% of tasks performed by humans in 2020 have been automated. Examples of this shift include:

- **Manufacturing:** Factories equipped with self-repairing robots operate 24/7, eliminating downtime. These robots assemble products with efficiency far surpassing human

- **Finance:** AGI systems analyze global markets in real-time, executing trades with unprecedented accuracy. Predictive algorithms manage individual portfolios, tailoring investments to unique risk profiles and goals.
- **Healthcare:** Robotic surgeons powered by AGI perform complex surgeries with unparalleled precision. Remote diagnostic systems bring top-tier healthcare to remote regions, closing the gap between urban and rural medical access.

Universal Basic Income (UBI)

As machines handle the bulk of economic production, **Universal Basic Income (UBI)** has become a cornerstone of modern economies, replacing wages as the primary means of income distribution. UBI ensures that every citizen receives a baseline income, providing financial security and reducing inequality. Early experiments in the 2020s, such as those in Finland and Kenya, paved the way for widespread adoption, demonstrating that UBI improves mental health, fosters entrepreneurship, and reduces financial stress.

Examples of UBI Implementation

1. **Kenya's GiveDirectly Experiment (2020s):**
   Participants in rural communities reported higher levels of economic activity and improved health outcomes after receiving direct cash transfers. By the 2030s, this model had expanded to several African nations, showcasing UBI's scalability.
2. **Finland (2017-2018):**
   Finland's pilot program revealed that recipients of a fixed income experienced reduced stress and greater satisfaction, even if employment rates remained stable.
3. **South Korea's Youth Dividend:**

This initiative provided regular payments to young people, funding education, entrepreneurship, and personal development. It fostered a generation of innovators and reduced youth unemployment.

UBI programs have allowed humanity to transition from labor-focused societies to purpose-driven ones. Freed from economic survival, individuals now focus on creativity, education, and personal fulfillment.

As Elon Musk predicted:
"There's a good chance we end up with universal basic income because of automation. The bigger question will be: What does humanity do next?"

2.3 Society Transformed: A Post-Work Economy

The automation of work has prompted a rethinking of cultural norms, values, and priorities. Without the need to toil for survival, society has shifted focus to well-being, lifelong learning, and community building.

Redefining Work and Purpose

Cultures historically tied personal identity and self-worth to labor. In a post-work world, this ethos has been redefined. Societies like Denmark, which already prioritized well-being over productivity in the early 21st century, have become global models. Here, success is measured not by income or hours worked, but by happiness, creativity, and the strength of social bonds.

Strategies for Adaptation

1. **Lifelong Learning:**
   Education systems emphasize adaptability, teaching skills like critical thinking, collaboration, and emotional intelligence—traits that complement, rather than

compete with, machines. Universities now focus on interdisciplinary approaches, blending the humanities with technology.

2. **Cultural Shifts:**
   Arts, philosophy, and community leadership have taken on renewed importance. Governments and organizations fund public art projects, collaborative innovation hubs, and programs to preserve cultural heritage in the face of globalization.
3. **Reinvention of Family and Leisure:**
   With increased free time, families spend more time together, fostering stronger bonds. Parents participate more actively in their children's education, while community engagement initiatives bring neighbors together.

2.4 Automation's Impact on Global Politics

The widespread adoption of automation and artificial intelligence has fundamentally altered the global political landscape by 2045. The traditional drivers of power—natural resources, manufacturing capacity, and labor force size—have been supplemented, and in many cases replaced, by technological prowess. Automation has not only widened the gap between nations that can leverage advanced technologies and those that cannot, but it has also introduced new tools for governance, competition, and even conflict.

This transformation has brought immense opportunities for progress but also heightened geopolitical tensions, as countries race to establish dominance in an increasingly AI-driven world.

Shifting Economic Power

Automation has redefined global economic hierarchies, with nations that embrace and lead in AI technologies solidifying their dominance over others. By 2045, economic power is concentrated in countries that have invested heavily in automation infrastructure, research, and talent development. These nations are no longer reliant on cheap labor, natural resources, or even trade agreements to drive growth. Instead, their economies are powered by automated industries, AGI-driven services, and self-sufficient production systems.

- **Repatriation of Manufacturing:**
  Advanced robotics and 3D printing technologies have enabled wealthy nations to bring manufacturing back home, bypassing the need for outsourcing to low-cost labor markets. This shift has disrupted traditional trade dynamics, leaving countries that once relied on exports of manufactured goods, like Bangladesh or Vietnam, scrambling to reinvent their economies.
- **AI-Driven Market Dominance:**
  Nations leading in AGI development—such as the United States, China, and India—control the most valuable global technologies, from automated supply chains to precision agriculture. These countries hold a strategic advantage in global markets, allowing them to dictate trade terms and influence international economic policies.

This economic realignment has created stark divisions between technology-rich and technology-poor nations, exacerbating inequality on a global scale. Developing countries that cannot compete in the AI economy face a widening "tech gap," which threatens to entrench them in cycles of poverty and dependence. Bridging this gap has become a central challenge for international organizations like the United Nations, which advocate for equitable technology sharing.

Geopolitical Competition: The AI Arms Race

By 2045, artificial intelligence and automation have become the centerpiece of geopolitical competition, often referred to as the "AI arms race." Nations view AI not only as a tool for economic growth but also as a strategic weapon in global conflicts. The development of AI-driven military technologies has become a defining feature of global power struggles, with advanced nations vying for dominance in autonomous systems, cybersecurity, and intelligence.

- **Autonomous Weapon Systems:**
  The militarization of AI has resulted in the deployment of autonomous drones, robotic soldiers, and precision strike systems capable of executing missions without human intervention. While these technologies reduce casualties for the deploying nation, they raise ethical and security concerns about accountability in war. For example, questions about how AGI-controlled weapons decide on targets and whether those decisions align with international law remain hotly debated.
- **Cybersecurity and AI Espionage:**
  Nations leverage AGI systems for cyberwarfare, conducting sophisticated hacking campaigns against rival states. AI systems can infiltrate secure networks, manipulate information, and disrupt critical infrastructure, such as power grids or communication systems. For instance, in 2043, a coordinated AI-driven cyberattack disabled the energy infrastructure of several small nations in Eastern Europe, sparking debates over international regulations for cyber warfare.
- **Surveillance and Control:**
  Authoritarian regimes have weaponized AI for mass surveillance, using facial recognition, predictive analytics, and social scoring systems to suppress dissent and control populations. These systems, modeled after China's Social Credit System in the 2020s, have been adopted by other nations, creating a global divide between democratic and authoritarian uses of AI.

The international community faces mounting pressure to regulate the use of AI in warfare and surveillance. Treaties akin to the Geneva Conventions have been proposed, aimed at defining ethical standards for autonomous weapon systems and outlawing their use in certain contexts. However, progress is slow, as nations are reluctant to cede their technological advantages.

Trade and Labor Realignments

Automation's impact on global labor markets has upended traditional trade systems. By 2045, low-cost labor is no longer a competitive advantage in international trade. Countries that once thrived as manufacturing hubs for wealthier nations have seen demand for their labor collapse, forcing them to reorient their economies toward technology-based industries or resource extraction.

- **Localized Manufacturing with Automation:**
  Advanced automation and 3D printing have allowed countries to produce goods domestically, reducing reliance on global supply chains. For example, the United States and Germany now produce 80% of their consumer goods using fully automated factories, a stark contrast to the reliance on Asian manufacturing in the early 21st century.
- **Resource Scarcity and Trade Wars:**
  While automation reduces dependence on labor, it increases demand for rare earth minerals and other raw materials essential for AI hardware and robotics. Countries with abundant mineral resources, such as the Democratic Republic of Congo, have become key players in a new era of resource-driven trade wars. The scramble for control over these resources has heightened tensions between global powers, particularly in regions like Africa and South America.

This disruption has also forced international organizations to rethink trade agreements. The World Trade Organization (WTO) now focuses on regulating access to AI intellectual property, rare earth minerals, and automation technologies, rather than traditional goods and services.

Global Governance in the Age of Automation

The rapid advancement of automation has exposed the limitations of existing global governance systems. International bodies struggle to keep pace with the ethical, economic, and security challenges posed by AI and robotics. However, by 2045, collaborative initiatives have emerged to address these gaps:

- **The AI Accord (2037):**
  Modeled after the Paris Climate Agreement, the AI Accord established international guidelines for the ethical use of AI. Participating nations agreed to ban autonomous weapons targeting civilians and to share AI research for humanitarian purposes, such as disaster relief and climate modeling.
- **Technology Equity Funds:**
  To address the growing tech gap between developed and developing nations, organizations like the World Bank and the African Union have established technology equity funds. These initiatives aim to provide underprivileged nations with access to AI infrastructure, training programs, and research collaborations, ensuring they are not left behind.
- **Digital Borders and Data Sovereignty:**
  Nations have increasingly asserted control over their digital borders, enacting laws to regulate the flow of data across international lines. These regulations, designed to protect citizens' privacy and national security, have created tensions between global tech companies and governments. For instance, the European Union's "Digital Sovereignty Act" limits the use of foreign AI systems, prioritizing local development.

Opportunities for Collaboration

Despite these tensions, automation also offers unprecedented opportunities for global cooperation. AGI has proven instrumental in solving problems that transcend borders, such as climate change, pandemics, and food security. For example:

- **Global Health Initiatives:**
  AI systems developed by international partnerships like the World Health Organization (WHO) and OpenAI analyze global health data to predict and prevent pandemics. In 2040, an AGI-driven early warning system helped contain a potential global outbreak of a novel virus, saving millions of lives.
- **AI for Peacekeeping:**
  The United Nations deploys AGI to monitor conflict zones, using drones and analytics to prevent violence before it escalates. By 2045, peacekeeping missions rely heavily on predictive models that identify flashpoints and recommend interventions.
- **Climate Collaboration:**
  Automation has revolutionized climate policy, enabling nations to jointly manage global emissions. Shared AI platforms monitor deforestation, ocean health, and carbon levels, providing real-time data for coordinated action.

Conclusion: The Future of Global Politics

Automation has profoundly reshaped the balance of global power, creating opportunities for collaboration while intensifying competition. By 2045, nations that embrace automation responsibly and equitably will emerge as leaders, while those that fail to adapt risk marginalization. The path forward depends on global cooperation, ethical governance, and a shared commitment to ensuring that technology serves humanity rather than divides it.

As automation continues to redefine global politics, the question for nations is no longer whether to adopt these technologies but *how* to wield them wisely.

2.5 Automation's Impact on Climate Policy

By 2045, automation has become one of humanity's most effective tools for combating climate change. Advanced artificial intelligence (AI), robotics, and machine learning have enabled precise management of resources, the acceleration of renewable energy, and large-scale environmental restoration. These systems do not just optimize existing methods; they redefine the possibilities of what humanity can achieve in its fight against environmental degradation. However, automation's potential as a climate solution is matched by significant challenges, from technological accessibility to ethical resource usage.

Optimizing Resource Management

Automation has redefined how resources are allocated, used, and monitored. AI systems provide real-time insights into global resource consumption, ensuring minimal waste while maximizing sustainability. These technologies have been especially transformative in agriculture, where advanced AI-driven tools ensure precision in farming practices.

- **Automated Irrigation Systems:** In arid regions, AI-powered irrigation systems equipped with soil sensors and predictive algorithms monitor moisture levels and weather conditions to deliver water only where and when it is needed. In Sub-Saharan Africa, these systems have reduced water waste by up to 60%, helping farmers adapt to prolonged droughts caused by climate change (UN FAO Report, 2040).
- **Energy Optimization in Urban Centers:** In 2045, "smart cities" powered by AI manage their energy consumption with precision. Buildings autonomously adjust heating, cooling, and lighting based on occupancy, while transportation systems dynamically route traffic to minimize congestion and fuel usage. Cities like Tokyo and Amsterdam have achieved net-zero emissions through these innovations, setting benchmarks for urban sustainability.

Accelerating Renewable Energy Adoption

The global transition from fossil fuels to renewable energy has been supercharged by automation. AI and robotics have removed many of the barriers to scaling renewable energy infrastructure, making it more efficient, cost-effective, and reliable than ever before.

- **Dynamic Optimization of Renewable Energy Systems:**
  By 2045, AI systems are the backbone of renewable energy grids, using advanced weather prediction models to maximize the output of wind farms and solar panels. For instance, AI platforms monitor wind patterns in real time, adjusting turbine angles to optimize power generation. In the Sahara Solar Belt, a sprawling network of solar farms powers much of Africa and Europe, managed entirely by automated systems that predict energy demands and adapt output accordingly.
- **Automated Maintenance Robots:**
  One of the greatest challenges for renewable energy infrastructure has historically been maintenance. Today, autonomous robots equipped with AI perform regular inspections and repairs on wind turbines, solar panels, and hydroelectric dams. These robots operate in harsh environments, reducing downtime and extending the lifespan of renewable energy systems.
- **Battery Storage Breakthroughs:**
  Advanced AI systems now oversee large-scale energy storage solutions, ensuring that renewable power is effectively stored and distributed. By 2045, most nations rely on AI-optimized battery farms, such as Tesla's Gigapack systems, to store excess solar and wind energy for use during peak demand periods.

Supporting Climate Modeling and Forecasting

Climate science has long depended on accurate modeling to understand and address environmental challenges. By 2045, AI systems have revolutionized this field, offering highly precise predictions that enable policymakers to act proactively rather than reactively.

- **Predictive Climate Models:**
  AI-powered climate models analyze vast datasets, including atmospheric CO_2 levels, ocean temperatures, and deforestation rates, to predict environmental changes with unparalleled accuracy. For example, NASA's Earth AI System provides 50-year projections of sea level rises and extreme weather patterns, giving governments a roadmap for proactive adaptation.
- **Early Warning Systems for Extreme Weather:**
  AI-driven systems monitor global weather in real time, issuing alerts for hurricanes, tsunamis, and heatwaves. In 2042, an AI-based weather monitoring platform prevented catastrophic flooding in Bangladesh by issuing precise evacuation orders 72 hours before a record-breaking monsoon.

Enabling Circular Economies

Automation has become central to the development of **circular economies**, which minimize waste by reusing and recycling resources. These systems reduce humanity's environmental footprint while creating sustainable cycles of production and consumption.

- **Automated Recycling Facilities:**
  AI-powered robots equipped with advanced vision systems and sensors sort waste with unparalleled accuracy, separating plastics, metals, and biodegradable materials. In South Korea, fully automated recycling facilities process 95% of the country's waste, drastically reducing landfill

usage.

- **Material Reuse in Manufacturing:**
  Factories now operate on closed-loop systems, where waste materials are recycled directly back into production lines. AI systems analyze materials during manufacturing to ensure maximum recyclability. For example, auto manufacturers use 3D printing techniques to construct vehicles from 80% recycled materials, dramatically reducing resource extraction.

Reforestation and Biodiversity Restoration

Reforestation and biodiversity conservation efforts have been revolutionized by robotics and AI, enabling humanity to combat deforestation and restore ecosystems at unprecedented scales.

- **Autonomous Reforestation Drones:**
  By 2045, drones are sowing billions of seeds annually, targeting areas devastated by deforestation or wildfires. Flash Forest (Canada) and BioCarbon Engineering (UK) have developed AI-driven drones that plant trees 10 times faster than manual methods. These drones analyze soil health and biodiversity to ensure that native species are replanted, maximizing ecological recovery.
- **Combatting Deforestation with AI:**
  AI technologies address deforestation at its roots. Satellite-based monitoring systems, such as Global Forest Watch, use machine learning to detect illegal logging activities in real time. Predictive analytics identify high-risk areas for deforestation, allowing governments to take preventative measures. Pachama, an AI-powered platform, helps companies reduce deforestation in their supply chains by tracking the environmental impact of their products.

Challenges and Opportunities

While automation has revolutionized climate policy, challenges remain.

- **Environmental Cost of Automation:**
  The production and disposal of automated systems, including robotics and batteries, can have negative environmental impacts if not managed sustainably. For example, the mining of rare earth minerals for AI hardware poses ecological risks. Governments must prioritize the development of green technologies that minimize these harms.
- **Equity in Access:**
  Advanced automation technologies are concentrated in wealthier nations, leaving many developing countries without access to these solutions. This disparity threatens to widen global inequalities in climate resilience, making international collaboration essential.
- **Unintended Consequences:**
  AI systems, while powerful, are not immune to errors. For instance, over-reliance on AI for resource management could lead to unforeseen disruptions, such as misallocation during extreme weather events. Human oversight remains critical to ensuring these systems function effectively.

A Collaborative Path Forward

To fully realize automation's potential in addressing climate change, nations must work together to create an inclusive and sustainable future.

- **International Technology Sharing:**
  Initiatives like AI for Earth and the Paris Agreement's technology-sharing framework have demonstrated the importance of global collaboration. These programs provide developing countries with access to AI tools, training, and infrastructure, ensuring that climate solutions are equitably distributed.
- **Ethical Development of Automation:**

Governments, corporations, and NGOs must establish clear ethical guidelines for the development and deployment of automation technologies. These guidelines should prioritize transparency, equity, and long-term sustainability.
- **Investing in Green AI:**
Research into "green AI" focuses on creating environmentally friendly algorithms and hardware. By designing energy-efficient systems, nations can reduce the ecological footprint of automation itself, ensuring that climate solutions do not inadvertently harm the planet.

Automation represents humanity's most powerful tool in the fight against climate change. By harnessing its potential responsibly, we can create a future that balances technological progress with environmental stewardship.

2.6 AI in Disaster Relief

By 2045, artificial intelligence has revolutionized disaster relief, enabling humanity to predict, prepare for, and respond to emergencies with unparalleled precision. AI systems, combined with automation and real-time data analysis, enhance every stage of disaster management—from early warning systems to post-disaster recovery. These tools are critical in a world increasingly shaped by climate-related disasters, urbanization, and global interconnectivity. The integration of AI into disaster relief is saving millions of lives, reducing economic losses, and fostering more resilient communities.

Predicting Disasters: From Uncertainty to Preparedness

The ability to predict disasters accurately has always been a cornerstone of effective disaster management. By 2045, AI-driven models analyze vast and complex datasets—including weather patterns, geological activity, ocean temperatures, and historical data—to predict natural disasters with near-perfect accuracy.

- **Early Warning Systems:**

AI-powered early warning systems issue real-time alerts for hurricanes, earthquakes, tsunamis, and more. IBM's Watson Weather Intelligence, for example, integrates data from satellites, ocean buoys, and seismic sensors to predict extreme weather events with a lead time of up to two weeks. In 2041, Watson accurately forecasted Hurricane Kassandra's path 10 days in advance, enabling governments across the Caribbean to evacuate 3 million people and prevent widespread loss of life.

- **Wildfire Monitoring:**
AI has become an essential tool in combating wildfires, one of the most destructive climate-related disasters. Systems like Google's AI Wildfire Initiative and Cal Fire's automated monitoring platform analyze satellite imagery, weather data, and vegetation density to detect wildfires within minutes of ignition. These systems predict the fire's spread, allowing firefighting teams to deploy resources to critical areas. In California, AI-driven wildfire detection systems have reduced the average response time by 75%, preventing billions of dollars in damage.

- **Earthquake Prediction Models:**
Although earthquakes have historically been challenging to predict, breakthroughs in AI-based geophysical modeling are improving accuracy. AI systems use data from seismic sensors and fault-line monitoring to identify subtle precursors to major earthquakes. In Japan, the QuakeNet AI platform correctly identified the 2042 Tokyo earthquake's epicenter 48 hours in advance, prompting the evacuation of high-risk zones and saving tens of thousands of lives.

Coordinating Emergency Responses: Precision Under Pressure

When disasters strike, the effectiveness of emergency response depends on the speed and accuracy of resource allocation. AI systems excel at processing vast amounts of information in real time, helping governments and relief organizations prioritize efforts where they are needed most.

- **Optimizing Resource Allocation:**
  AI-driven platforms analyze population density, infrastructure damage, and supply chain logistics to ensure that resources such as food, water, medical supplies, and shelter reach affected areas efficiently. For instance, the United Nations World Food Programme's AI logistics system has cut delivery times by half during crises, ensuring critical supplies arrive within 24 hours of a disaster.
- **AI-Powered Drones in Disaster Zones:**
  Drones equipped with AI are transforming how relief operations assess and access disaster-stricken areas. High-resolution cameras and thermal sensors allow drones to map affected regions, locate survivors, and identify damaged infrastructure. In remote areas devastated by the 2043 Philippines typhoon, AI-guided drones delivered emergency medical supplies to isolated communities unreachable by traditional means.
- **Real-Time Communication and Translation:**
  In multilingual regions, communication during disasters can be a bottleneck for effective response. AI-powered chatbots and real-time translation tools, such as Microsoft Translator, facilitate clear communication between aid organizations, governments, and local communities. During the 2040 South Asian floods, these tools enabled international relief teams to coordinate with local authorities, bridging language barriers and speeding up relief efforts.

Post-Disaster Recovery: Rebuilding Smarter and Faster

AI systems are critical not only in immediate disaster response but also in the long-term recovery process. By streamlining damage assessments, guiding infrastructure rebuilding, and fostering resilience, AI ensures that communities recover faster and stronger.

- **Damage Assessment:**
AI-powered platforms analyze satellite imagery and drone data to assess damage to roads, buildings, and utilities. Companies like Planet Labs use machine learning to generate detailed, real-time maps of disaster-hit areas, enabling governments to prioritize repairs and allocate resources effectively. After the 2044 Istanbul earthquake, AI assessments reduced the time required for damage evaluation from weeks to hours, accelerating recovery efforts.
- **Designing Resilient Infrastructure:**
AI systems simulate disaster scenarios and optimize construction methods to create disaster-resilient infrastructure. In coastal regions prone to hurricanes, for example, AI-guided urban planning incorporates flood-resistant materials and elevated structures, ensuring cities can withstand future events. Singapore's AI Urban Design Lab has become a global leader in this field, rebuilding neighborhoods destroyed by typhoons with structures that are 90% more resilient to wind and flooding.
- **Supporting Mental Health Recovery:**
AI-powered mental health platforms assist survivors in coping with trauma after disasters. Virtual counselors, such as those developed by MindEase AI, provide personalized support and monitor users' progress, ensuring access to mental health resources in areas where human counselors are scarce. This has significantly reduced long-term psychological impacts among survivors in disaster-prone regions.

AI-Driven Collaboration Across Stakeholders

Effective disaster relief relies on coordination between governments, non-governmental organizations (NGOs), and private sector partners. AI acts as a bridge between these stakeholders, providing shared platforms for data analysis, communication, and decision-making.

- **The Global Disaster Alert and Coordination System (GDACS):**
  By 2045, GDACS integrates AI-powered dashboards that provide real-time updates on disaster severity, resource availability, and humanitarian needs. This system enables stakeholders to collaborate seamlessly, ensuring a unified and efficient response.
- **AI in Cross-Border Disasters:**
  Disasters often transcend national borders, requiring international coordination. AI-driven models predict cascading effects, such as how floods in one region might disrupt food supplies in another. For example, during the 2042 Mekong River floods, AI platforms helped Vietnam, Cambodia, and Laos coordinate efforts, sharing resources to prevent food shortages and water contamination across the region.

Challenges and Ethical Considerations

While AI offers groundbreaking solutions for disaster relief, it also introduces unique challenges that must be addressed to ensure equitable and ethical outcomes.

1. **Data Gaps and Inequality:**
   AI systems rely on vast datasets, but underdeveloped regions often lack the infrastructure to collect and share this data. Without inclusive data collection, disaster relief efforts risk prioritizing wealthier, data-rich areas over marginalized communities.
2. **Algorithmic Bias:**
   AI models may inadvertently prioritize certain regions or demographics based on biases in their training data. For instance, densely populated urban areas might

receive more attention than rural communities, even when both face similar risks. Oversight mechanisms are essential to ensure equitable resource distribution.

3. **Privacy Concerns:**
   During disasters, the collection of personal data, such as geolocation and health records, raises privacy concerns. Governments and organizations must strike a balance between effective relief efforts and safeguarding individual rights, particularly in regions with limited data protection laws.

4. **Dependence on AI Systems:**
   Over-reliance on AI systems can create vulnerabilities, particularly if these systems fail or are compromised. For example, a cyberattack targeting disaster relief algorithms could disrupt critical operations, highlighting the need for robust cybersecurity measures and human oversight.

The Future of AI in Disaster Relief

Looking ahead, the integration of AI into disaster relief will continue to evolve, offering even more advanced capabilities:

- **Predictive Ecosystem Models:**
  By 2050, AI systems will predict not just single disasters but interconnected events, such as how droughts might trigger wildfires or how hurricanes might cause power grid failures. These predictive ecosystems will allow governments to prepare for cascading crises proactively.

- **Fully Autonomous Relief Operations:**
  The next frontier in disaster relief is fully autonomous systems that can handle complex tasks without human intervention. AI-powered robotic teams could simultaneously extinguish wildfires, repair infrastructure, and deliver supplies, functioning as an all-in-one disaster response unit.

- **Resilience through Simulation:**
  Communities will increasingly use AI simulations to model potential disasters and test their preparedness. Virtual reality training platforms powered by AI will immerse policymakers, emergency responders, and citizens in lifelike disaster scenarios, improving coordination and readiness.

Conclusion: Redefining Relief with AI

Artificial intelligence has transformed disaster relief into a proactive, data-driven field capable of saving lives, reducing suffering, and rebuilding communities with unprecedented speed and precision. However, as with any powerful tool, AI must be wielded responsibly, ensuring that its benefits are shared equitably across all regions and demographics. By addressing the ethical challenges of AI deployment and fostering international collaboration, humanity can harness this technology to create a safer and more resilient future.

CHAPTER 3: SOCIETY IN TRANSITION

Humanity is standing at a pivotal moment in history. As AI and automation reshape industries, economies, and cultures, society finds itself in the midst of profound transformation. This transition is about more than just adapting to new technologies—it's about redefining what it means to be human in a world where machines shoulder more of our responsibilities and even begin to share in our creativity. How do we navigate this shift while preserving the essence of human values, creativity, and connection? How do we ensure that technological progress uplifts everyone, rather than dividing us further?

This chapter explores how society is evolving in response to automation, addressing the challenges of integration, the transformation of education, the reshaping of cultural and social structures, and the redefinition of human purpose.

3.1 Bridging the Gap: Technology and Humanity

The rapid pace of technological advancement has created a world of unprecedented convenience, efficiency, and possibility. However, it has also introduced significant challenges in maintaining a balance between technological integration and human-centric values. How do we ensure that technology complements humanity rather than overshadowing it? Bridging this gap is one of the defining challenges of the 21st century.

Challenges of Integration

1. **The Digital Divide**
   Despite advances in AI and automation, not everyone benefits equally. The **digital divide**—the gap between those who have access to advanced technology and those who do not—has widened disparities in education, healthcare, and economic opportunities. Rural areas, underserved communities, and developing nations often lack reliable internet access, digital infrastructure, or the resources to adopt new technologies.
   - **Educational Disparities:** In many regions, students are unable to participate in online learning due to limited internet connectivity or the lack of affordable devices. Meanwhile, children in urban areas benefit from AI-driven personalized education systems that adapt to their needs. Bridging this gap requires substantial investment in broadband expansion, affordable technology programs, and government-backed digital literacy initiatives.
   - **Healthcare Gaps:** Telemedicine and AI-powered diagnostics have revolutionized healthcare in wealthier areas, but communities without access to these technologies are left behind. Programs like Google's "Project Taara," which delivers affordable internet to remote regions via laser technology, provide hope for bridging these divides.
   - **Global Efforts:** Countries like Estonia, often called the "world's first digital nation," have created national digital strategies to ensure equitable access to technology. Estonia's e-residency program, for example, has empowered entrepreneurs worldwide to participate in its digital economy, demonstrating the power of inclusive technological frameworks.
2. **The Erosion of Privacy**

As AI systems monitor behaviors and collect personal data, concerns about privacy have skyrocketed. Wearable health trackers, smart home devices, and even social media platforms gather sensitive information about users, raising the risk of breaches and misuse. For example, smart refrigerators in some regions have been hacked to access data about household routines, and facial recognition technologies have been used for unauthorized surveillance.

Solutions:
- Governments must enact robust **data protection laws**, such as the EU's **General Data Protection Regulation (GDPR)**, which mandates transparency and user consent in data collection.
- Companies should adopt ethical frameworks that prioritize user privacy, implement encryption for sensitive data, and ensure transparency about how personal information is used.
- Education on personal cybersecurity practices, such as controlling app permissions and using secure networks, is also essential to empower individuals.

3. **The Loss of Human-Centric Roles**

Automation is replacing jobs across industries, leaving many individuals struggling to find new roles that provide a sense of purpose and identity. Factory workers displaced by robotics and customer service professionals replaced by AI chatbots are examples of the disruptive nature of this transition.

Solutions:
- Governments and private organizations must invest in **reskilling initiatives** to prepare workers for jobs in emerging fields, such as AI ethics, green energy, and creative industries.
- Programs like Germany's **Dual Vocational Training System**, which combines education with practical work experience, provide a successful model for equipping

workers with future-ready skills.
- Societies must redefine what it means to contribute meaningfully, encouraging pursuits in creativity, caregiving, and community-building, which cannot be automated.

Opportunities for Harmony

1. **Tech-Enhanced Creativity**
   Automation has the potential to free humans from repetitive tasks, allowing them to focus on creative pursuits that amplify their ingenuity.
   - **Generative AI in the Arts:** Tools like Adobe Firefly and DeepArt enable artists to blend technology with traditional mediums, pushing the boundaries of creative expression.
   - **Music and Literature:** Musicians use platforms like OpenAI's MuseNet to compose symphonies, while authors collaborate with AI to craft novels that explore complex ideas.
   - **Invention and Innovation:** Designers and inventors rely on machine learning algorithms to prototype products and simulate designs, reducing the time needed to bring ideas to life.
2. **Human-AI Collaboration**
   Rather than competing with machines, humans are increasingly working alongside AI to solve complex problems.
   - **Healthcare Innovations:** AI accelerates research in genomics and drug discovery, identifying patterns that human researchers might overlook. For example, DeepMind's AlphaFold has revolutionized protein structure prediction, unlocking breakthroughs in treating diseases.
   - **Architecture and Design:** AI simulates structural integrity and energy efficiency, while architects

focus on aesthetic and cultural elements, creating a seamless partnership between human intuition and computational precision.
3. **Building Ethical Frameworks**
Embedding ethics into AI development ensures that technology evolves in alignment with human values.
   - **Explainable AI:** Ensuring that AI decisions are transparent and understandable builds public trust. For instance, IBM's Explainability 360 toolkit helps developers clarify how algorithms make predictions.
   - **Bias Mitigation:** Companies like OpenAI are auditing their systems to identify and reduce biases, ensuring fair and equitable outcomes across diverse populations.
   - **Global Collaboration:** Initiatives like UNESCO's **Recommendation on the Ethics of Artificial Intelligence** are fostering international cooperation to set ethical standards for AI deployment.

3.2 Redefining Education for the Future

As society transitions into an era dominated by automation and AI, education must adapt to prepare individuals for a landscape where traditional skills are no longer enough. This transformation requires reimagining education as a lifelong journey, emphasizing adaptability, creativity, and emotional intelligence.

Adaptive Learning

AI-powered platforms now offer **personalized education experiences**, tailoring lessons to each student's unique needs and learning pace.

- **Virtual Reality (VR) and Augmented Reality (AR):** Immersive learning environments allow students to explore concepts in vivid detail. For example, biology students can conduct virtual dissections, and history students can "walk"

through ancient cities using AR headsets.
- **Global Access to Education:** Platforms like Khan Academy, coupled with AI-driven tutors, provide high-quality education to students in remote regions, bridging gaps in access and quality.

Lifelong Learning

The rapid evolution of industries requires individuals to continuously update their skills throughout their lives.

- **Corporate Reskilling Programs:** Companies like Amazon and Microsoft have launched initiatives to retrain employees for roles in emerging fields, ensuring their workforce remains competitive.
- **Online Learning Platforms:** Platforms like Coursera, Udemy, and edX offer flexible courses in AI, data science, and digital marketing, empowering individuals to learn at their own pace.

Emphasis on Soft Skills

In a world where technical skills can be automated, soft skills such as **critical thinking, creativity, emotional intelligence, and adaptability** are becoming invaluable.

- **Future-Focused Curriculums:** Schools are incorporating project-based learning and interdisciplinary approaches to foster problem-solving and innovation.
- **Emotional Intelligence Training:** Programs that teach empathy, leadership, and collaboration prepare students to thrive in workplaces increasingly driven by teamwork and human-AI interaction.

3.3 Social Structures in Flux

Automation and AI have transformed the fabric of social and cultural dynamics, reshaping how individuals interact within families, communities, and across cultural boundaries.

Changing Family Dynamics

Automation has redistributed time, freeing families from many household chores and work obligations.

- **Parenting in the Digital Age:** Parents must balance the benefits of educational apps and devices with the risks of overexposure to screens. AI-powered parenting assistants, such as conversational robots that read stories or tutor children, provide support but cannot replace meaningful human interactions.
- **Strengthened Bonds:** With more free time, families are investing in quality time, such as shared hobbies, travel, or community engagement.

Evolving Community Bonds

Technology has transformed how people connect and interact, creating both opportunities and challenges for community-building.

- **Digital Communities:** Social media platforms and virtual spaces have allowed individuals to form global connections, but these communities can sometimes foster echo chambers or misinformation.
- **Reviving Local Interactions:** The localization of production and services, driven by automation, has renewed interest in local initiatives such as farmers' markets and makerspaces, fostering face-to-face interactions and community resilience.

3.4 The Human Spirit: Resilience and Reinvention

Despite the disruptions caused by automation and AI, humanity has always shown resilience in the face of change. By embracing reinvention, individuals and societies can turn challenges into opportunities.

Redefining Human Purpose

In an automated world, human roles will focus on creativity, connection, and exploration:

1. **Creators and Innovators:** Artists, designers, and inventors will continue to push boundaries, blending human intuition with AI capabilities.
2. **Ethical Stewards:** As technology evolves, roles in ensuring alignment with human values—such as AI ethicists and policy advisors—will be critical.
3. **Explorers:** Freed from repetitive work, humanity can focus on exploring new frontiers, from space colonization to the mysteries of the deep sea.

Emerging Future Careers

New careers will emerge as societies adapt to automation, including:

- **Virtual Experience Designers**
- **Sustainability Advocates**
- **AI Collaboration Specialists**
- **Longevity Specialists**

Conclusion: Celebrating Humanity

As machines take over repetitive tasks, humans have the opportunity to focus on what makes them unique: empathy, creativity, and the ability to dream. By leveraging these qualities, societies can ensure that technology complements rather than overshadows the human experience, building a future where innovation and humanity thrive hand in hand.

CHAPTER 4: THE FUTURE OF WORK IN AN AUTOMATED WORLD

The rise of AI and automation has transformed the nature of work, redefining productivity, purpose, and the very concept of employment. These advancements have triggered significant shifts across industries, cultures, and economies. While many traditional jobs are disappearing, new opportunities are emerging, forcing societies to adapt and reimagine the meaning of work in a post-automation era.

This chapter explores the changes reshaping industries, the challenges of job displacement, the emergence of Universal Basic Income (UBI), and the opportunities for redefining human purpose and fulfillment in a world where machines do the heavy lifting.

4.1 The Shift in Employment Paradigms

AI and automation are fundamentally altering the global employment landscape, driving a shift from human-centered labor to hybrid and fully automated systems. This transformation is not merely about job displacement—it is about reimagining the workforce and identifying new areas where human potential can thrive alongside technological advancements.

Transforming Industries: A Sector-by-Sector Evolution

Automation is reshaping nearly every sector, from healthcare to entertainment, introducing efficiencies, innovations, and disruptions. Below is a closer look at how industries are transforming in the age of automation:

1. **Healthcare:**
   - **AI Applications:** AI-powered diagnostic tools, such as those developed by Google Health and IBM Watson, analyze medical imaging with greater accuracy than human doctors. Predictive analytics help identify at-risk populations, while robotic surgeries like those performed by the Da Vinci system ensure precision and reduce recovery times.
   - **Personalized Medicine:** AGI systems analyze genetic data and patient histories to deliver highly customized treatment plans, reducing trial-and-error approaches in medicine.
2. **Finance:**
   - **Algorithmic Trading:** AI handles billions of trades daily, identifying patterns humans might miss, while robo-advisors manage personalized investment portfolios for everyday consumers.
   - **Fraud Detection:** Real-time AI systems monitor financial transactions for anomalies, protecting institutions and customers from cybercrime.
3. **Retail and E-commerce:**
   - **Customer Personalization:** AI recommendation engines curate shopping experiences, increasing customer satisfaction and driving sales.
   - **Logistics Automation:** Robots in Amazon warehouses handle inventory management, while cashier-less stores like Amazon Go redefine convenience shopping.
4. **Transportation and Logistics:**

- **Autonomous Vehicles:** Self-driving trucks powered by companies like Tesla and Waymo optimize long-haul logistics, while drones revolutionize last-mile deliveries.
- **Real-Time Tracking:** AI-powered systems provide instant visibility into supply chains, enabling just-in-time inventory management.

5. **Agriculture:**
    - **Precision Farming:** Drones equipped with AI sensors monitor crop health, soil conditions, and water usage, enabling resource-efficient farming.
    - **Automated Equipment:** Autonomous tractors and harvesters reduce labor costs and improve productivity.
6. **Entertainment:**
    - **AI-Generated Content:** AI algorithms create music, art, and video content, while personalization engines recommend content tailored to individual preferences.
    - **Virtual Production:** The film industry increasingly relies on AI for special effects, virtual actors, and script analysis, lowering production costs while enhancing creativity.
7. **Education:**
    - **AI Tutors:** Tools like Duolingo and Khan Academy adapt lessons to individual students' learning styles, ensuring maximum engagement and comprehension.
    - **Immersive Learning:** Virtual reality (VR) and augmented reality (AR) transform classrooms into interactive, hands-on environments.

Industries Facing Limited Benefits from AI

While automation enhances efficiency in many sectors, certain industries thrive on human creativity, emotional intelligence, and cultural specificity, making them less conducive to full automation.

1. **Artisan Crafts and Handicrafts:**
   Handcrafted goods hold cultural and aesthetic value that cannot be replicated by machines. For example, bespoke furniture, jewelry, and textiles retain their uniqueness because of their imperfections and the personal touch of the craftsman.
2. **Performing Arts:**
   Live performances in theater, dance, and music rely on human spontaneity and emotional connection with the audience. AI-generated performances may complement, but cannot replace, this authenticity.
3. **Religious and Spiritual Guidance:**
   Faith-based leadership and community rituals depend on human empathy, trust, and cultural traditions, which are deeply personal and often resistant to technological intervention.
4. **Psychotherapy and Counseling:**
   While AI chatbots like Woebot provide initial support, the trust, nuance, and deep listening required for effective therapy cannot be fully automated.

Industries Resisting AI Integration

Certain industries actively resist AI due to ethical, cultural, or operational concerns:

- **Education (Traditional Models):** Despite the proliferation of AI tutors, traditional teaching emphasizes mentorship and face-to-face interactions. Many educators argue for the irreplaceable role of human guidance in nurturing critical thinking and emotional growth.
- **Hospitality (Boutique Services):** Luxury hotels, small bed-and-breakfasts, and high-end dining rely on personalized human interactions to create memorable experiences.

4.2 Automation and Job Displacement

Automation is transforming industries, but its impact on jobs is uneven. While it reduces costs and improves efficiency, it also displaces millions of workers who previously held routine and repetitive roles.

Industries Experiencing Early Job Loss

1. **Manufacturing:** Assembly-line jobs are being replaced by robots capable of working 24/7 with greater precision and fewer errors.
2. **Transportation and Logistics:** Autonomous trucks, drones, and automated warehouses are disrupting traditional roles like truck drivers, delivery workers, and inventory managers.
3. **Retail:** Self-checkout systems and cashier-less stores are replacing retail staff, while e-commerce platforms reduce the need for in-store roles.
4. **Food Service:** Fast food outlets increasingly use robots to cook and serve meals, reducing the need for kitchen staff and cashiers.

Emergence of New Roles

Despite the displacement of traditional jobs, automation is creating entirely new career paths in technology, ethics, and sustainability. Emerging roles include:

- **AI Trainers:** Professionals who teach AI systems to interpret complex human behavior and emotions.
- **Ethical Technologists:** Specialists who ensure AI systems align with societal values and regulatory standards.
- **Sustainability Scientists:** Innovators who use AI to combat climate change through renewable energy, conservation, and waste reduction.

Critical Skills in a Post-AI World

To thrive in the age of automation, individuals must cultivate uniquely human skills that complement AI systems:

1. **Emotional Intelligence:** Leadership, conflict resolution, and caregiving roles rely heavily on the ability to understand and manage emotions.
2. **Critical Thinking and Creativity:** Innovation and problem-solving remain vital as machines handle routine analysis.
3. **Adaptability and Resilience:** Continuous learning ensures workers can evolve with technological advances.

4.3 Universal Basic Income and Beyond

The concept of **Universal Basic Income (UBI)** has gained traction as a solution to automation-driven economic disruption. UBI provides a guaranteed income to all citizens, offering financial security as traditional jobs become scarce.

Pilot Programs and Success Stories

1. **Finland (2017-2018):** Unemployed citizens received €560 monthly, reducing stress and improving well-being.
2. **Kenya (Ongoing):** Nonprofits provide basic income to rural communities, enabling investments in education, small businesses, and health.

Broader Implications of UBI

1. **Encouraging Innovation:** Freed from financial stress, individuals can pursue creative projects and entrepreneurial ventures.
2. **Improved Health Outcomes:** Studies link guaranteed income to reduced anxiety, improved nutrition, and

better access to preventive care.

4.4 Redefining Purpose in a Post-Work Society

As work becomes less central to human identity, societies are exploring new avenues for fulfillment, focusing on passion projects, volunteerism, and lifelong learning.

The Rise of Passion Projects

Freed from monotonous labor, people have more time to invest in creative and educational pursuits. Examples include:

- Developing independent art, music, or writing projects.
- Contributing to open-source tech initiatives or local charities.

Lifelong Learning and Exploration

With access to platforms like Coursera and Khan Academy, individuals can continuously acquire skills, explore new subjects, and maintain intellectual curiosity.

4.5 Envisioning the Future of Work

In an automated world, work becomes a blend of creativity, collaboration, and purpose-driven efforts. The future of work is not defined by jobs lost but by opportunities gained in reimagining human potential. By embracing adaptability, fostering inclusivity, and prioritizing ethical innovation, societies can thrive in the age of automation.

CHAPTER 5: AI'S ROLE IN RESHAPING SOCIETY

The rise of artificial intelligence has been a defining moment in the evolution of human society. Its influence extends beyond technical innovation, fundamentally altering how people interact, make decisions, and shape their future. As AI becomes embedded in every facet of daily life, its role in reshaping society brings both transformative opportunities and urgent challenges. From redefining human relationships to reshaping cities, tackling global challenges, and raising philosophical questions about the human experience, AI's societal influence is as profound as it is far-reaching.

5.1 Redefining Human Interaction

AI is reshaping how humans connect, communicate, and interact, both with one another and with machines. While technology has always played a role in shaping communication, the advent of AI-driven tools has introduced unprecedented levels of personalization, automation, and complexity to these interactions.

AI in Communication

1. **Language Translation:**
   AI-powered tools such as Google Translate, DeepL,

and Microsoft Translator have bridged global communication barriers, allowing people from different linguistic backgrounds to interact effortlessly. These tools have revolutionized cross-border commerce, diplomacy, and social relationships. For example, real-time AI-powered translators enable international business meetings without the need for human interpreters.

- **Opportunities:**
  - These tools expand access to education, commerce, and cultural exchange, fostering inclusivity in an increasingly globalized world.
- **Challenges:**
  - Critics argue that AI translators often struggle with cultural nuances, idiomatic expressions, or context-specific meanings, sometimes leading to misunderstandings.

2. **Virtual Assistants and AI Chatbots:**
AI assistants like Siri, Alexa, and Google Assistant have become integral to daily life, helping users manage schedules, search for information, and even control smart homes. Meanwhile, AI chatbots like ChatGPT facilitate human-like conversations for both personal and business purposes.

- **Impact on Society:**
  - These systems enhance convenience but raise questions about dependency. As humans grow accustomed to interacting with AI systems for tasks like customer service or personal organization, interpersonal communication may decline, affecting social dynamics.
  - Example: Japan's increasing use of AI-powered robots for elder care has sparked debates about replacing human caregivers with machines, even though these robots provide companionship and practical support.

Social Media Algorithms and AI's Role in Shaping Conversations

AI powers content curation and recommendation engines on platforms like Facebook, YouTube, TikTok, and Instagram, influencing how people consume information and connect with others.

1. **Echo Chambers and Polarization:**
   Social media algorithms tailor content based on user behavior, preferences, and past interactions. While this personalization improves user engagement, it also risks creating echo chambers that reinforce biases and limit exposure to diverse perspectives.
   o **Real-World Impact:**
   During the 2016 U.S. election, AI-driven algorithms on social media platforms were criticized for amplifying misinformation and fueling political polarization. Similarly, studies have shown that algorithms can amplify extremist content by prioritizing sensationalism over accuracy.
2. **Combatting Misinformation:**
   On the flip side, AI tools are being developed to identify and flag false information. Platforms like Twitter and YouTube use machine learning to monitor posts for misinformation, though balancing free speech with moderation remains a contentious issue.

5.2 Ethics in AI Governance

The rapid adoption of AI has outpaced the creation of ethical and regulatory frameworks to govern its development and application. As AI systems make more autonomous decisions, ensuring they align with societal values and ethical principles is paramount.

Accountability and Transparency

1. **Explainable AI (XAI):**
   AI systems often operate as "black boxes," making decisions without human operators fully understanding how or why. Explainable AI (XAI) seeks to address this by creating systems that can articulate the rationale behind their decisions in human terms.
   o **Example:** The healthcare industry increasingly demands XAI in AI diagnostic tools to explain why a system flagged a potential cancer diagnosis, ensuring trust and accountability among doctors and patients.
2. **Liability in Autonomous Systems:**
   As AI takes on roles traditionally performed by humans, questions of accountability arise. For example, if an autonomous vehicle causes an accident, who is responsible—the manufacturer, the developer, or the operator? Governments and legal systems worldwide are grappling with these questions.

Bias and Fairness in AI

Bias in AI systems reflects the biases present in the data on which they are trained. Without careful intervention, these systems can perpetuate or even amplify existing inequalities.

1. **Examples of Bias:**
   o   Hiring algorithms have been found to favor male candidates for tech jobs based on historical data that reflected male-dominated workforces.
   o   Facial recognition software has been criticized for higher error rates when identifying people of color, leading to concerns about discriminatory policing.
2. **Solutions:**
   o   Diverse development teams, rigorous bias audits, and the use of balanced, representative datasets are critical

to building fair AI systems. Initiatives like IBM's AI Fairness 360 toolkit and Microsoft's FairLearn platform aim to address these issues.

5.3 Redesigning Cities with AI

AI is transforming urban planning and development, helping cities become smarter, more sustainable, and more resilient in the face of growing populations and climate change.

Smart Infrastructure

1. **Transportation:**
   o   AI-driven traffic management systems analyze real-time data to reduce congestion and optimize public transit. Autonomous vehicles powered by AI further enhance urban mobility by improving safety and accessibility.
   o   Example: In Singapore, the AI-powered smart traffic system uses predictive algorithms to adjust traffic lights, reducing travel times by 20%.
2. **Energy Efficiency:**
   o Smart grids leverage AI to optimize energy distribution, integrating renewable sources like solar and wind while minimizing waste. Cities like Amsterdam and Copenhagen have implemented AI systems to manage energy usage and improve sustainability.
3. **Public Services:**
   o   Predictive maintenance powered by AI reduces costs and improves service delivery. For instance, AI sensors in water systems can detect leaks early, preventing waste and reducing repair costs.

Sustainability Goals

AI supports sustainability initiatives through innovations in urban farming, waste management, and water conservation.

1. **Urban Farming:**
   - AI-driven vertical farms use machine learning to optimize crop yields in small urban spaces. Companies like AeroFarms are revolutionizing agriculture by producing fresh, pesticide-free food using 95% less water.
2. **Waste Management:**
   - Smart waste bins equipped with AI sensors sort recyclables and reduce landfill waste. Municipalities in South Korea and Japan have adopted these technologies to achieve zero-waste targets.

5.4 AI's Role in Global Challenges

AI holds immense potential to tackle some of the world's most pressing issues, from climate change to healthcare and economic inequality.

Climate Change Mitigation

1. **Predictive Analytics for Environmental Risks:**
   - AI systems model deforestation, rising sea levels, and natural disaster risks, empowering governments to take proactive measures. NASA uses AI to predict wildfire behavior and allocate firefighting resources efficiently.
2. **Optimizing Renewable Energy:**
   - AI algorithms optimize the performance of wind turbines and solar panels, maximizing energy output and minimizing downtime. AI also forecasts weather patterns to align renewable energy production with demand.

Healthcare Access

AI-driven tools bridge healthcare gaps in underserved areas by providing affordable and accessible medical support.

1. **Telemedicine:**
   - AI chatbots like Babylon Health assist patients with preliminary diagnoses and health advice, reducing strain on healthcare systems.
2. **Global Vaccine Distribution:**
   - AI systems optimize vaccine distribution, ensuring equitable access in remote and economically disadvantaged areas.

5.5 Education and Empowerment

Education is the foundation of societal progress, and AI has emerged as a transformative force in making high-quality learning accessible to people worldwide. Through adaptive learning systems, immersive technologies, and enhanced accessibility tools, AI is empowering individuals across all demographics to learn, grow, and innovate. By tailoring education to individual needs and eliminating barriers to access, AI is not only democratizing education but also redefining its purpose in a world increasingly shaped by automation.

Personalized Learning for All

One of AI's most revolutionary contributions to education is the ability to provide **personalized learning experiences**. Unlike traditional one-size-fits-all models, AI-powered platforms adapt to each learner's unique strengths, weaknesses, and preferences.

1. **AI-Driven Adaptive Platforms:**
   - Tools like DreamBox, Coursera, and Byju's analyze how students engage with content and adjust lessons in real-time to optimize comprehension and retention. For example, if a student struggles with algebraic concepts, the platform may revisit foundational principles or present the material in a different format, such as

interactive videos or step-by-step tutorials.
- Example: In rural Kenya, students using the onebillion learning app, powered by AI, showed literacy and numeracy gains equivalent to an additional year of schooling in just nine months.

2. **Learning Beyond the Classroom:**
- AI is making education accessible outside traditional settings. Platforms like Khan Academy and Google Classroom allow students to learn at their own pace and revisit topics as needed.
- AI chatbots act as tutors, providing real-time feedback and explanations. For example, Squirrel AI in China has improved math and science performance by offering students tailored learning paths and consistent support.

3. **Bridging Skill Gaps:**
- For professionals, AI platforms identify gaps in knowledge and provide microlearning modules tailored to their career goals. Platforms like LinkedIn Learning use AI to suggest courses based on a user's skills, career trajectory, and market trends, helping individuals remain competitive in rapidly evolving industries.

Making Education Accessible to Underserved Communities

AI is breaking down barriers to education for populations that have historically been excluded due to socioeconomic, geographic, or physical limitations.

1. **Global Access Through Online Learning:**
- AI-enabled platforms are bringing high-quality education to remote areas. For example, UNICEF partnered with Microsoft's AI for Accessibility to create **Learning Passport**, a platform that provides education to displaced and underserved children, including refugees.

- Organizations like Worldreader use AI to distribute e-books to children in low-income countries, ensuring access to literacy resources even in areas without physical libraries.

2. **Accessibility for People with Disabilities:**
    - AI-powered tools like speech-to-text systems, screen readers, and sign-language recognition software are making education inclusive for students with disabilities. For example, Microsoft's Seeing AI app enables visually impaired students to interact with written material through audio descriptions.
    - Tools like Otter.ai provide real-time transcription for hearing-impaired students in classrooms, improving inclusivity.

3. **Affordable Learning Tools:**
    - AI-powered education systems, often accessible through smartphones, reduce the cost of learning. Apps like Duolingo democratize language education, while low-cost virtual tutoring powered by AI reduces reliance on expensive private tutors.

Lifelong Learning and Workforce Development

In an age where automation rapidly transforms industries, education is no longer confined to childhood or young adulthood. AI is enabling **lifelong learning**, equipping individuals to adapt to changing job markets and personal aspirations.

1. **Continuous Upskilling for Evolving Careers:**
    - Platforms like Skillsoft and edX partner with companies to offer industry-specific certifications powered by AI. These programs analyze workforce needs and create targeted courses to prepare employees for future roles.
    - Example: Google's Career Certificates program leverages AI to teach job-ready skills in high-demand

fields like IT support, data analytics, and UX design, enabling individuals to transition into new careers quickly.
2. **Democratizing Higher Education:**
   o   AI enables modular, flexible degree programs that cater to working adults. Universities like Arizona State University and Georgia Tech use AI tools to track student performance, predict challenges, and recommend resources, increasing graduation rates for non-traditional learners.
3. **Career Guidance and Job Matching:**
   o   AI-powered career counseling systems, like Joblift and Pymetrics, analyze a person's skills, interests, and experiences to recommend career paths or match them with employers. By assessing soft skills and behavioral traits, these systems identify opportunities that humans might overlook.
4. **Customized Corporate Training:**
   o   Companies increasingly rely on AI-powered training systems to prepare their workforce for digital transformation. AI analyzes employee performance data to create customized development plans, fostering productivity and engagement.

The Role of AI in Redefining Educational Content

AI is not only delivering education but also **creating new forms of content** tailored to modern learners.

1. **Interactive and Immersive Experiences:**
   o   Virtual reality (VR) and augmented reality (AR) powered by AI are revolutionizing experiential learning. Students can now conduct virtual science experiments, explore historical landmarks, or simulate workplace scenarios. For example, Google Expeditions allows students to take virtual field trips to places like

the Great Wall of China or the surface of Mars.
- AI-generated simulations help medical students practice surgeries or allow engineering students to design and test structures in virtual environments.

2. **Real-Time Updates in Learning Materials:**
- AI systems can instantly update digital textbooks and course materials to reflect the latest research and developments in any field, ensuring students always learn from up-to-date information.

3. **Language Localization:**
- AI-driven translation tools ensure that educational content is available in multiple languages, making global knowledge accessible to diverse populations. Khan Academy's content is now translated into over 40 languages using AI tools, expanding its reach to millions of students worldwide.

Addressing Challenges in AI-Driven Education

While AI offers transformative opportunities, it also raises challenges that must be addressed to ensure education remains equitable and effective.

1. **Digital Divide:**
- Unequal access to technology remains a barrier. Without investments in broadband infrastructure and affordable devices, students in rural or low-income areas risk being left behind.
- Solutions: Public-private partnerships, such as the United Nations' **Giga Project** (which aims to connect every school in the world to the internet), can help bridge this divide.

2. **Data Privacy and Security:**
- AI systems often rely on collecting vast amounts of student data, raising concerns about privacy and misuse.

- o   Solutions: Governments must enforce strict data protection laws, such as the GDPR in Europe, while schools and platforms adopt secure, ethical data practices.
3. **Over-Reliance on Technology:**
- o   While AI enhances learning, excessive dependence on technology risks diminishing critical thinking, creativity, and interpersonal skills.
- o   Solutions: Hybrid learning models that integrate human mentorship alongside AI systems ensure a balance between technological and human instruction.
4. **Bias in AI Systems:**
- o   If improperly trained, AI systems can reflect biases in the data they are built on, reinforcing stereotypes or creating inequitable learning experiences.
- o   Solutions: Diverse datasets, regular audits, and inclusive design teams are essential to eliminating bias in educational AI systems.

Empowering Educators Through AI

AI is not a replacement for teachers but a powerful tool to enhance their effectiveness and free them from administrative burdens, allowing them to focus on personalized support for students.

1. **Streamlining Administrative Tasks:**
- o   AI tools automate grading, attendance tracking, and performance analysis, enabling teachers to dedicate more time to instruction and mentorship.
2. **Professional Development:**
- o   AI-driven platforms recommend personalized training programs for educators, helping them stay updated on new teaching methods and technologies.
3. **Supporting Differentiated Instruction:**
- o   AI systems provide teachers with real-time insights into student performance, allowing them to

identify struggling students and tailor interventions accordingly.

Conclusion: AI as a Catalyst for Global Learning

AI's integration into education has the potential to bridge social divides, unlock individual potential, and prepare societies for the challenges of the future. By addressing barriers to access, ensuring ethical implementation, and balancing technology with human connection, AI can empower learners worldwide to thrive in a rapidly changing world. As we navigate this transformation, the goal must remain clear: to create a more inclusive, innovative, and equitable educational system that leaves no one behind.

Personalized Learning: Tailoring Education for Every Learner

Traditional education has long followed a "one-size-fits-all" model, but AI is disrupting this paradigm by enabling learning experiences that adapt to the needs, pace, and preferences of individual students. Personalized learning powered by AI is providing tailored educational experiences for students of all ages and backgrounds, ensuring that everyone—whether a child in a remote village or an executive seeking career advancement—has access to tools that cater to their unique needs.

The Mechanics of AI-Driven Personalized Learning

AI-powered personalized learning platforms use advanced algorithms, data analysis, and machine learning to adjust content delivery based on each learner's progress, strengths, weaknesses, and preferences. These systems continuously assess how students engage with material, offering instant feedback and adapting the curriculum dynamically to maximize learning outcomes.

- **Dynamic Adaptation:** AI tracks a student's performance in real time, detecting areas of struggle and areas of mastery. For example, if a student repeatedly struggles with fractions in a math module, the system can pause progression, introduce simpler foundational concepts, or offer alternative teaching approaches (such as visual aids, practice exercises, or gamified content) to reinforce understanding.
- **Predictive Analytics:** AI-powered platforms predict where students may face challenges based on their past performance, offering preemptive interventions. For instance, if a student shows a pattern of disengagement with a particular topic, the system might introduce motivational tools like progress rewards or adjust the difficulty level of the material to re-engage them.
- **Behavioral Data Integration:** AI doesn't just track academic performance; it can monitor behavioral patterns, such as how long a student spends on each task, their response to challenges, or their engagement with specific types of content. This insight allows the platform to refine its approach over time, creating a continuously optimized learning experience.

Real-World Examples of Personalized Learning Platforms

Several AI-powered platforms are already revolutionizing personalized learning globally:

1. **DreamBox Learning (K-8 Education):**
   DreamBox is an adaptive math platform that tailors lessons to elementary and middle school students. The platform analyzes over 50,000 data points per hour as students engage with content, customizing lessons to meet individual needs. It encourages self-paced learning, allowing advanced students to progress faster while providing struggling students with additional support.

2. **Duolingo (Language Learning):**
   With over 500 million users, Duolingo leverages AI to tailor language lessons to individual learners. The app assesses how well a user has mastered vocabulary and grammar concepts, adjusting difficulty levels and introducing new material at an optimal pace. Gamification elements, like streaks and rewards, further motivate users to continue learning.
3. **Squirrel AI (China):**
   This AI-driven tutoring system focuses on STEM education and has been implemented in schools across China. It breaks complex topics into micro-lessons, tailoring instruction for each student. Squirrel AI has achieved impressive results, with students showing up to a 30% improvement in test scores compared to traditional methods.
4. **Carnegie Learning (High School and Beyond):**
   This platform uses AI to create personalized math pathways for high school and college students. Carnegie Learning's adaptive technology adjusts lessons in real time and provides teachers with actionable insights about student performance, enabling more effective intervention in the classroom.

The Impact of Personalized Learning on Education Systems

The widespread adoption of personalized learning is reshaping the goals and structure of education systems worldwide. By focusing on individual growth and mastery, these platforms are delivering several key benefits:

1. **Closing Achievement Gaps:**
   Personalized learning platforms address the unique needs of underserved students, such as those in low-income or rural communities, who may lack access to high-quality teachers or resources. By adapting to each

learner's needs, AI-powered systems ensure that no one is left behind.

- **Case Study:** A UNICEF initiative partnered with Century Tech, an AI-powered personalized learning platform, to provide education to refugee children in Lebanon. The system identified each child's starting level in literacy and math and adjusted lessons to help them catch up, resulting in significant academic progress within six months.

2. **Accelerating Advanced Learners:**
Students who excel in certain subjects are no longer constrained by the pace of a traditional classroom. AI-powered systems allow advanced learners to explore more challenging material at their own pace, fostering a deeper understanding and preventing boredom.

- Example: In the U.S., students using the Zearn platform for math showed faster progression, with gifted learners advancing through material twice as quickly as their peers in traditional classroom settings.

3. **Fostering Self-Paced Learning:**
Personalized learning empowers students to take control of their education, allowing them to progress at their own pace without feeling pressured or held back by their peers. This is particularly beneficial for adult learners and working professionals juggling education with other responsibilities.

4. **Improving Teacher Effectiveness:**
AI-powered systems provide teachers with actionable insights into student performance, enabling them to intervene where needed most. Teachers can focus on mentoring, creative instruction, and fostering critical thinking skills while AI handles routine tasks like grading and performance tracking.

Expanding Access Through Multilingual Personalization

One of the most transformative impacts of AI-driven personalized learning is its ability to serve diverse populations through **multilingual education tools**.

- Platforms like Duolingo and Google's AI-Powered Read Along provide lessons in multiple languages, enabling non-native speakers to learn effectively.
- In India, AI education startup ConveGenius provides personalized content in multiple regional languages, helping bridge learning gaps for underserved students in rural areas.

Personalized Learning in Higher Education and Corporate Training

The role of AI in personalized learning extends far beyond primary and secondary education:

1. **In Higher Education:**
   Universities like Arizona State University and Georgia Tech are implementing AI to enhance retention rates and reduce dropout rates. For instance, ASU's AI-powered platform, **EAB Navigate**, identifies at-risk students based on their engagement patterns and academic performance, offering personalized support before problems escalate.
2. **In Corporate Training:**
   AI-powered training programs are reshaping workforce development by tailoring learning modules to employees' current skills and career goals. Companies like IBM and PwC use AI platforms to deliver personalized upskilling, ensuring that workers are prepared for digital transformation.

Challenges of Personalized Learning

While the potential of AI-driven personalized learning is immense, several challenges must be addressed to fully realize its benefits:

1. **Digital Divide:**
   Access to personalized learning platforms depends on reliable internet and affordable devices, which are not universally available. Bridging this gap requires significant investment in digital infrastructure and affordable technology.
2. **Data Privacy Concerns:**
   Personalized learning relies on vast amounts of student data to tailor experiences, raising questions about how this data is stored, shared, and protected. Strong data governance frameworks, such as Europe's GDPR, are critical to ensuring student privacy.
3. **Over-Reliance on Technology:**
   There's a risk that education systems may overemphasize AI tools at the expense of human interaction, which is essential for fostering emotional intelligence, collaboration, and social skills. Hybrid models combining AI with teacher-guided instruction are vital to address this concern.
4. **Algorithmic Bias:**
   If the AI models behind personalized learning platforms are trained on biased datasets, they risk perpetuating inequities in education. Regular audits and inclusive training data are necessary to ensure fairness.

The Future of Personalized Learning

The future of education lies in integrating personalized learning with emerging technologies like virtual reality, augmented reality, and even the metaverse. Imagine classrooms where students can "step into" a historical event, conduct simulated lab experiments, or explore mathematical concepts in 3D—all tailored to their individual learning paths.

In addition, AI will enable cross-cultural collaboration through virtual classrooms, where students from different countries can learn together in real time, breaking down barriers and fostering global connections.

Conclusion: Personalized Learning as a Catalyst for Change

AI-powered personalized learning is revolutionizing education, making it more inclusive, engaging, and effective. By tailoring content to the needs of individual learners, these systems are breaking down traditional barriers and unlocking human potential on an unprecedented scale. As society addresses challenges like the digital divide and data privacy, personalized learning will continue to transform lives, ensuring that education is not just a privilege but a universal right.

CHAPTER 6: A VISION FOR 2045 AND BEYOND

6.1 The AI-Human Collaboration

As we look toward 2045, the future is defined by the integration of AI as a collaborative partner in society. Rather than replacing human roles, AI enhances creativity, efficiency, and problem-solving across industries.

Synergistic Roles

- **Co-Creation in Art and Design:** AI tools will continue to assist artists and designers in producing innovative work, blending human intuition with computational precision.
- **Scientific Discovery:** AI-powered simulations and analysis will accelerate breakthroughs in fields like medicine, physics, and environmental science.

Ethical Collaboration

- Ensuring AI aligns with human values requires ongoing vigilance. Collaborative partnerships between technologists, ethicists, and policymakers will define frameworks for responsible AI use.

6.2 Envisioning Future Societal Structures

AI's transformative potential extends beyond industries, influencing how societies organize and interact.

Decentralized Governance

- AI-enabled decision-making could make governments more efficient and transparent. Decentralized models powered by blockchain and AI could increase citizen participation in policymaking.

Redefining Community

- Virtual spaces and AI-driven platforms may create new forms of communities based on shared interests rather than geography, fostering global collaboration.

6.3 Education and Knowledge in the AI Era

The future of education revolves around preparing individuals for a post-work society.

AI as a Mentor

- AI tutors will provide personalized education, guiding learners through adaptive curriculums tailored to their needs.
- Virtual and augmented reality experiences will immerse students in interactive learning environments, making education more engaging and accessible.

Lifelong Learning Networks

- Knowledge-sharing platforms and AI-facilitated peer-to-peer mentorships will emphasize continuous learning, ensuring individuals adapt to technological advancements.

6.4 The Role of Philosophy and Ethics

The integration of AI raises fundamental questions about the essence of humanity and the direction of societal progress.

Rethinking Purpose

- As AI reduces the necessity for traditional labor, societies will explore philosophical concepts of purpose, fulfillment, and identity.

- Cultural and spiritual frameworks may evolve to address the psychological impact of an AI-driven world.

Balancing Progress

- Ethical boundaries will need to evolve alongside technological progress, ensuring humanity retains control over AI's trajectory.

6.5 Global Collaboration for the AI Age

AI's impact is universal, demanding coordinated efforts to address shared challenges and opportunities.

International Standards

- Developing global guidelines for AI ethics, privacy, and security will ensure equitable and responsible AI deployment worldwide.

Bridging the Digital Divide

- Ensuring access to AI tools and education for underserved communities will reduce inequalities and foster inclusive innovation.

6.6 AI's Impact on Mental Health

AI has the potential to transform mental health care by improving access, personalization, and early intervention, but it also introduces new challenges that require careful consideration.

Enhancing Mental Health Services

1. **AI-Powered Therapy Tools:**
   - Chatbots like Woebot and Wysa provide accessible mental health support, offering cognitive behavioral therapy (CBT) techniques and emotional guidance. These tools can act as a first line of defense, particularly in regions with limited access to professional care.
2. **Personalized Interventions:**
   - AI analyzes individual data to tailor mental health

treatments, identifying patterns in mood, behavior, and sleep. This personalization enhances the effectiveness of interventions and reduces the trial-and-error period in traditional therapy.

3. **Early Detection:**
   - AI-driven analytics identify early warning signs of mental health conditions, such as depression or anxiety, by monitoring digital footprints, wearable devices, or voice patterns. Early detection allows for timely intervention, improving outcomes.

4. **Virtual Reality Therapy:**
   - VR experiences powered by AI create immersive environments for exposure therapy and relaxation techniques, effectively treating conditions like PTSD, phobias, and stress.

Challenges and Ethical Concerns

1. **Privacy Risks:**
   - The collection of sensitive mental health data by AI systems raises concerns about privacy and misuse. Robust encryption and ethical data practices are essential to maintaining trust.

2. **Over-Reliance on Technology:**
   - While AI tools are valuable, they should not replace human therapists. The lack of empathy and nuanced understanding in AI systems limits their ability to address complex emotional needs.

3. **Bias in Algorithms:**
   - AI systems trained on non-representative data may provide inaccurate or harmful recommendations for underrepresented populations. Ensuring diverse datasets is crucial for equitable mental health care.

Opportunities for Progress

1. **Global Accessibility:**
   - AI-powered mental health tools democratize care,

reaching underserved populations and reducing stigma associated with seeking help.

2. **Support for Professionals:**
   - AI systems assist therapists by streamlining administrative tasks and providing insights, allowing clinicians to focus more on patient care.

3. **Community Integration:**
   - AI can foster mental health awareness through interactive platforms, promoting community-driven support networks and peer counseling programs.

By addressing challenges and embracing innovation, AI can significantly enhance mental health care while ensuring ethical and equitable implementation.

6.7 Challenges and Opportunities Ahead

The journey to 2045 is fraught with challenges but brimming with potential.

Anticipating Risks

- From AI bias to environmental impacts, proactive measures will be essential to address unintended consequences of AI advancements.

Harnessing Opportunities

- By embracing collaboration, innovation, and ethical stewardship, societies can unlock the transformative potential of AI, ensuring a brighter future for all.

CHAPTER 7: THE GLOBAL FUTURE OF AI

The future of artificial intelligence is a global phenomenon, transcending borders and impacting humanity as a whole. By 2045, AI will not only drive innovation but also serve as a unifying force to tackle complex global challenges. As nations invest in AI, it will play a pivotal role in fostering collaboration, preserving cultural diversity, addressing climate change, and shaping international policies. However, navigating the geopolitical, ethical, and regulatory challenges associated with AI requires collective effort and a shared vision for humanity's progress.

7.1 AI as a Catalyst for Global Collaboration

AI has the power to unite nations, accelerate innovation, and tackle issues that no single country can address alone. By fostering collaboration in research, disaster response, and sustainable development, AI can build bridges where there were once divides.

Shared Scientific Advancements

AI's unparalleled ability to analyze massive datasets and uncover patterns is revolutionizing global research, fostering international partnerships, and accelerating breakthroughs that benefit humanity as a whole.

1. **Accelerating Healthcare Innovation:**

- During the COVID-19 pandemic, AI-powered platforms like BlueDot identified and tracked outbreaks early, enabling countries to coordinate responses. AI also played a critical role in vaccine development, with systems like IBM Watson analyzing virus structures to accelerate breakthroughs.
- **Future Potential:** By 2045, AI is expected to eliminate major diseases such as cancer and Alzheimer's through precision medicine and early detection. Global collaborations, fueled by AI, will enable universal access to affordable treatments.

2. **Tackling Climate Change:**
- AI-driven climate models allow researchers to predict the impact of environmental policies and identify solutions for mitigating global warming. For example, IBM's Green Horizon platform uses AI to forecast air pollution and recommend strategies for reducing emissions.
- By pooling data from satellite systems, ocean sensors, and agricultural reports, nations can collaborate on climate policies with unprecedented precision.

3. **Space Exploration:**
- The global AI community is driving a new era of space exploration. AI-powered telescopes and data analysis tools help scientists from different nations collaborate on missions to identify habitable planets, mine asteroids for rare resources, and understand cosmic phenomena.
- **Example:** NASA's AI partnerships with the European Space Agency (ESA) and private firms like SpaceX have fostered unprecedented cooperation in the search for extraterrestrial life and the development of lunar and Martian colonies.

AI Advancements Driving Global Collaboration

1. **Real-Time Translation Tools:**
   AI-powered systems like DeepL and Google Translate are eliminating language barriers, facilitating international diplomacy, trade negotiations, and cultural exchange. By 2045, AI translation tools will enable real-time communication in every language, creating a truly global society.
   o **Example:** An international summit on climate change could see delegates from over 100 countries communicating seamlessly using AI-powered headsets that translate speech in real time while preserving cultural nuance.
2. **Collaborative Research Platforms:**
   Platforms such as DeepMind's AlphaFold have already revolutionized fields like biology by solving protein-folding problems. In the future, AI-powered global research hubs will allow scientists to collaborate in real time, sharing findings and resources to tackle complex problems.
   o **Example:** An AI-powered open platform for disease research could connect scientists from Africa, Asia, and South America, enabling them to co-develop vaccines for regional epidemics while leveraging AI simulations.
3. **Disaster Response Coordination:**
   o AI systems integrate satellite imagery, weather data, and ground-level reports to provide real-time insights during natural disasters. This allows nations to coordinate rescue efforts and allocate resources more effectively.
   o **Example:** In a future scenario, AI-powered disaster response platforms could mobilize international aid during a global flooding crisis, providing drones and robots to deliver supplies, map affected regions, and ensure equitable distribution of resources.

Addressing Global Challenges with AI

1. **Supporting the UN's Sustainable Development Goals (SDGs):**
   AI is critical to achieving the UN's 17 SDGs, which include eradicating poverty, ensuring food security, and promoting sustainable energy.
   - **Example:** AI-powered agricultural systems, such as FarmSense, use drones and sensors to monitor soil quality, optimize irrigation, and predict crop yields, directly addressing global food security.
2. **Fostering Equity in Global Health Initiatives:**
   - AI helps bridge healthcare gaps in underserved regions by enabling telemedicine, mobile diagnostics, and predictive health analytics.
   - **Example:** By 2045, AI-powered health systems could bring affordable healthcare to billions of people in remote areas, integrating wearable sensors that detect early signs of disease and connect patients with virtual doctors.
3. **Smart Infrastructure for Global Cities:**
   - International partnerships are leveraging AI to build sustainable urban infrastructure, from energy-efficient grids to AI-driven traffic systems that reduce emissions and improve mobility.
   - **Example:** Cities like Singapore, Amsterdam, and Tokyo are collaborating on AI projects to create "green" smart cities, sharing lessons and technologies with developing nations to help them build sustainable urban centers.

7.2 Geopolitical Implications of AI

As AI reshapes global industries, it has become a key factor in the balance of international power. While AI presents opportunities for collaboration, it also introduces geopolitical tensions and raises ethical concerns about its use in defense and governance.

The AI Arms Race

1. **Strategic Advantages in Defense:**
   Nations are racing to develop AI-powered defense technologies, from autonomous drones to cybersecurity tools capable of neutralizing threats in real time.
   - **Example:** The U.S., China, and Russia are leading in AI military innovation, raising concerns about the potential misuse of autonomous weapons in conflicts.
2. **Economic Inequality and Technological Gaps:**
   - Advanced nations with significant AI investments are gaining disproportionate economic power, while developing countries risk falling further behind. Addressing this divide requires global collaboration and equitable AI access.

Ethical Leadership on the Global Stage

1. **Promoting Peaceful AI Use:**
   Leading AI nations bear a responsibility to promote ethical standards that ensure AI is used for peaceful and humanitarian purposes.
   - **Example:** A "Global AI Peace Accord" could be established to ban the use of AI in offensive military applications, akin to nuclear disarmament treaties.
2. **Ensuring Transparency in AI Use:**
   - Nations must collaborate to create transparent AI governance frameworks that prevent authoritarian regimes from exploiting AI for surveillance and oppression.

7.3 Balancing AI Innovation and Regulation

The rapid pace of AI innovation demands a careful balance

between encouraging technological progress and establishing regulatory safeguards to ensure ethical and equitable outcomes.

Harmonizing International Standards

1. **Global AI Ethics Frameworks:**
   - Organizations like UNESCO, the OECD, and the World Economic Forum are working to establish consistent global standards for AI ethics, data privacy, and security.
   - **Example:** By 2045, nations could adopt a unified global AI regulatory framework, ensuring all AI systems are tested for bias, transparency, and fairness before deployment.
2. **Preventing AI Misuse:**
   - International agreements must prohibit the use of AI for exploitative purposes, such as mass surveillance, predatory financial systems, or discriminatory hiring practices.

7.4 AI and Cultural Preservation

AI plays a critical role in preserving cultural diversity, ensuring that globalization does not erode humanity's rich tapestry of languages, traditions, and histories.

Reviving Languages

1. **Saving Endangered Tongues:**
   - AI-driven tools, such as Rosetta AI, document and revitalize endangered languages by creating digital archives, translation tools, and language-learning apps.
   - **Example:** By 2045, AI will have fully reconstructed extinct languages such as Ancient Egyptian, enabling historians to teach them to new generations.

Digitizing and Preserving Heritage

1. **Virtual Museums and Archives:**
   - AI digitizes cultural artifacts, creating virtual museums accessible to anyone with an internet connection.
   - **Example:** The Louvre collaborates with AI developers to create a global virtual museum where people worldwide can explore the world's artistic treasures in immersive 3D.

7.5 The Role of Education in a Global AI Era

Education is the cornerstone of a globally interconnected AI future. By 2045, education systems will emphasize cross-cultural competencies, digital literacy, and collaborative skills to prepare students for global challenges.

Digital Inclusion

1. **Bridging the Divide:**
   - AI-powered education platforms will ensure students in rural or developing regions have access to the same quality of education as their urban counterparts.
   - **Example:** Solar-powered AI education pods in remote villages provide students with access to global curricula, taught in their native languages.

7.6 A Shared Vision for the Future

The global future of AI is one of interconnectedness, shared progress, and collective responsibility. By working together to align AI development with human values, nations can create a future where technology uplifts every member of society.

Empowering Humanity

AI is not just a tool for progress but a force for equality, creativity, and global prosperity. Its future success depends on humanity's ability to balance ambition with ethical stewardship, ensuring that no one is left behind.

CHAPTER 8: PERSONAL FUTURES IN THE AGE OF AI

8.1 How AI Will Influence Daily Life

The influence of AI will be omnipresent—streamlining mundane tasks, optimizing personal choices, and enabling greater control over our environments. Adapting proactively to these changes will ensure individuals are not overwhelmed but empowered by AI's integration into their daily routines.

Personalized Assistance in Everyday Life

1. **Smart Homes of the Future:**
   By 2045, AI-powered smart homes will go far beyond today's voice-controlled devices. They will anticipate and adapt to individual needs without being explicitly told.
   - **Temperature and Lighting:** AI systems will monitor weather patterns, time of day, and your preferences to adjust heating, cooling, and lighting automatically for comfort and energy efficiency.
   - **Automated Grocery Management:** AI-powered refrigerators and pantries will track food inventory, suggest recipes based on available ingredients, and even place orders for groceries before you run out.
   - **Cleaning and Maintenance:** Robots equipped with

AI will clean homes, perform minor repairs, and even identify potential maintenance issues like plumbing leaks or electrical faults before they occur.
- **Example:** A smart home detects that you've had a stressful day (based on your tone during phone calls and wearable health data) and creates a calming evening by dimming the lights, playing relaxing music, and ordering your favorite comfort food.

2. **AI in Personal Health Management:**
Wearable devices powered by AI will continuously monitor health metrics like heart rate, sleep patterns, blood sugar levels, and even stress levels.
- **Predictive Health Monitoring:** These devices will identify early signs of illnesses, such as diabetes, heart disease, or depression, and suggest preventive measures.
- **Personalized Wellness Plans:** AI will analyze your fitness, nutrition, and mental health needs to create tailored wellness recommendations. For example, it might recommend yoga to lower stress levels or adjust your diet to combat fatigue.
- **Example:** By analyzing your data, your wearable device notices an irregular heart rhythm and alerts your doctor for a follow-up, potentially preventing a major health issue.

3. **AI as a Personal Advisor for Everyday Decisions:**
AI-powered virtual assistants will act as personal advisors for nearly every aspect of life.
- **Financial Planning:** AI will help manage budgets, suggest investments, and predict future expenses based on spending habits.
- **Meal Planning:** Virtual nutritionists will create meal plans based on your health goals, allergies, and even your cultural preferences.
- **Travel Assistance:** AI systems will monitor your calendar and preferences to suggest travel itineraries,

book flights, and recommend activities tailored to your interests.
- **Example:** AI notices your energy usage is higher than normal and suggests switching to a more affordable and sustainable energy plan, saving you money while reducing your carbon footprint.

8.2 Building Relationships with AI

As AI becomes more integrated into our personal lives, it will play a growing role in how we build and maintain relationships—both with other humans and with the AI systems themselves.

AI as a Companion

1. **Social Robots:**
   AI-powered robots will offer companionship, particularly for individuals who are lonely, elderly, or living in remote areas.
   - These robots will provide conversation, reminders for medication or appointments, and even emotional support.
   - **Example:** ElliQ, an AI companion for seniors, provides not only reminders but also proactive engagement, such as suggesting a walk when it detects inactivity or initiating games to keep the user mentally active.
2. **Virtual Friends:**
   Platforms like **Replika** already allow users to engage in meaningful conversations with AI-powered chatbots. By 2045, virtual companions will become even more sophisticated, simulating empathy, humor, and deep conversations.
   - **Future Scenario:** A person working abroad uses a virtual friend for emotional support, having daily conversations that help reduce loneliness and maintain mental well-being.

Ethical Considerations in AI Relationships

The rise of AI companions will also spark critical ethical and social questions:

- **Dependency on AI:** Could reliance on AI for emotional support reduce human-to-human connections?
- **Privacy Concerns:** How will AI companions handle sensitive personal data?
- **Authenticity:** Can an AI-powered relationship truly replace a human one?
- **Preparing Today:** Engage with early AI companions like virtual chatbots to explore their capabilities, benefits, and limitations, while maintaining awareness of these ethical dilemmas.

8.3 AI's Role in Lifelong Learning

The rapid pace of technological change means that learning will no longer be confined to childhood or formal education. AI will play a pivotal role in facilitating lifelong education and skill acquisition.

Adaptive Learning Platforms

1. **Hyper-Personalized Curricula:**
   AI systems will create learning experiences tailored to individual styles and goals.
   - **Example:** A graphic designer uses an AI-driven platform to learn coding in a way that aligns with their visual learning style, incorporating project-based lessons that simulate real-world design challenges.
2. **Gamification of Education:**
   AI will transform education into engaging, game-like experiences to encourage participation and retention.
   - **Example:** Learning a new language might involve

an AI-powered VR environment where users engage in real-time conversations with simulated native speakers.

Skills for a Post-AI World

1. **Human-Centric Skills:**
   AI will handle most routine tasks, shifting demand to uniquely human skills like creativity, adaptability, and emotional intelligence.
   - **Preparing Today:** Start focusing on skills that complement AI, such as collaborative problem-solving, design thinking, and leadership.
2. **Microlearning for Rapid Skill Acquisition:**
   AI-driven platforms like **Coursera** and **Khan Academy** will offer bite-sized lessons tailored to busy lifestyles, enabling people to stay updated in fast-evolving fields.

8.4 Personal Ethics in an AI World

As AI becomes more integrated into our lives, individuals will need to make deliberate ethical choices about how they use and interact with these systems.

Navigating Data Privacy

1. **Owning Your Data:**
   Individuals must become advocates for their data, ensuring it is used transparently and responsibly.
   - **Example:** Opt for AI systems that prioritize user privacy through encryption and data minimization, and regularly review your privacy settings.
2. **Balancing Convenience and Control:**
   While AI makes life easier, it is critical to ensure you remain the ultimate decision-maker in your life.
   - **Example:** Use AI tools to generate options or recommendations, but take the final decision yourself

to maintain autonomy.

8.5 AI and Personal Fulfillment

By automating mundane tasks, AI will free individuals to focus on activities that bring meaning, creativity, and joy to their lives.

Time for Creativity and Exploration

1. **Automating Mundane Chores:**
   With AI handling repetitive tasks, individuals will have more time to pursue hobbies, art, and personal passions.
   - **Example:** A musician uses AI to handle time-consuming music transcription, freeing more time for composing and performing.
2. **Rediscovering Purpose:**
   In a world where work is less central, individuals will have the opportunity to explore deeper questions of purpose, spirituality, and connection.

Strengthening Relationships Across Distances

AI will make maintaining relationships easier through advanced communication tools:

- **Example:** Real-time translation tools allow families and friends to maintain meaningful connections, even across language barriers.

8.6 Preparing for a Personal AI Future

The key to thriving in an AI-driven future lies in proactive preparation.

Cultivating Digital Literacy

1. **Understanding AI Systems:**

Take online courses or attend workshops to build foundational knowledge of how AI works.
- **Example:** Platforms like **edX** and **Udemy** offer accessible AI literacy courses for beginners.

Advocating for Ethical AI Development

1. **Becoming an AI Advocate:**
   Join local or online communities focused on AI ethics to participate in shaping how AI impacts society.

Conclusion

By 2045, AI will have transformed every facet of daily life. Those who adapt early by building skills, embracing lifelong learning, and making deliberate choices about how they integrate AI into their lives will thrive in this exciting new era. Proactive preparation today ensures that you'll not only navigate the changes AI brings but also seize the opportunities it creates to lead a more meaningful, fulfilling life.

FINAL THOUGHTS AND CONCLUSION: EMBRACING THE AGE OF AI

A New Dawn for Humanity

The era of artificial intelligence marks one of the most profound shifts in human history—a leap as monumental as the Industrial Revolution or the dawn of the Internet. It is a technological evolution that challenges not just what we do but who we are. AI is no longer a distant idea; it is embedded in our daily lives, from how we communicate to how we learn, work, and innovate. As this transformation accelerates, the question before us is not whether we will adapt, but how.

Will we passively allow AI to shape our world, or will we actively engage to shape AI in alignment with our highest values? The answer lies in our ability to think critically, act collaboratively, and embrace the opportunities AI offers while addressing its risks with wisdom and foresight.

The Power of Adaptation: A Story of Resilience

History shows that humanity's greatest strength lies in its adaptability. From fire to electricity, from steam engines to the Internet, every technological revolution has tested our capacity to innovate, grow, and transform. AI is no different. It represents both a challenge and an opportunity—a tool that can amplify our creativity, alleviate our burdens, and solve problems that once seemed insurmountable.

But adaptation is not automatic. It requires action, curiosity, and a commitment to learning. Those who act early, who develop the skills and mindsets to thrive in this new era, will not only adapt but flourish. They will not only benefit from AI but help shape its impact on the world.

- **Lessons from the Industrial Revolution:** The Industrial Revolution disrupted traditional agricultural economies and created new industries, professions, and ways of life. Those who resisted change struggled, while those who embraced it found new opportunities. Similarly, AI offers the chance to redefine work, education, and creativity for generations to come.
- **A Personal Call to Adapt:** Ask yourself today: How can I prepare for an AI-driven world? Whether it's learning a new skill, educating others, or engaging in ethical conversations about technology, every small step taken now contributes to a brighter, more inclusive future.

Addressing Public Concerns: AI and Society

As AI transforms society, it also brings fears about its potential misuse. Concerns surrounding privacy, surveillance, bias, and the loss of control over technology are valid—and urgent. Addressing these issues with transparency, accountability, and ethics will be critical to ensuring AI uplifts rather than oppresses.

Surveillance and Privacy: Striking the Balance

AI-powered surveillance systems, such as facial recognition, predictive policing, and mass data collection, have raised alarms about the erosion of privacy and civil liberties. These tools, while useful in enhancing security, risk creating authoritarian overreach or a "surveillance state" if left unchecked.

- **Challenges of Facial Recognition:**
  - **Bias and Accuracy:** Studies have shown that facial recognition systems are prone to racial and gender biases, leading to wrongful identifications and discrimination.
  - **Mass Surveillance:** The proliferation of cameras in public spaces risks normalizing constant surveillance, chilling free expression and eroding personal freedoms.
  - **Example:** In 2020, cities like San Francisco banned the use of facial recognition technology due to its potential misuse and lack of accountability. These examples show the need for careful regulation.
- **Ethical Strategies for Use:**
  - **Regulation First:** Governments must implement laws that define when, where, and how surveillance AI can be used, ensuring it does not infringe on rights.
  - **Public Consent:** Communities must be involved in deciding whether these systems are deployed, with transparency about how data will be used.
  - **Independent Audits:** Regular audits of AI systems must identify biases, improve accuracy, and hold developers accountable for ethical practices.

Predictive Policing and Judicial AI

Predictive policing tools use historical crime data to forecast future incidents, enabling law enforcement to allocate resources more efficiently. However, if the data reflects historical biases —such as over-policing in certain communities—AI risks perpetuating systemic inequities.

- **Risks in the Justice System:**
  - **Opaque Decision-Making:** AI algorithms are often "black boxes," meaning their logic is difficult to

understand or challenge.
- Bias in Sentencing Recommendations: AI tools have been shown to recommend harsher sentences for minorities due to biased training data.
- **Proposed Solutions:**
  - Explainable AI (XAI): AI systems must be transparent about how decisions are made, allowing for oversight and appeals.
  - Human Oversight: Judges, police officers, and other professionals must remain in control of decision-making, using AI only as a supplemental tool.

Military and Security AI

The use of AI in autonomous weapons and military applications poses profound ethical challenges. International agreements, akin to nuclear non-proliferation treaties, will be critical to preventing misuse or an AI arms race.

- **Example of Progress:** The United Nations has begun drafting guidelines for the use of lethal autonomous weapons, emphasizing human oversight and accountability.

Reimagining Purpose in an AI World

One of the most profound shifts AI will bring is the redefinition of human purpose. Freed from routine labor, humanity will have the opportunity to focus on creativity, relationships, and self-discovery. This is not a loss of meaning—it is a chance to expand it.

Rediscovering What It Means to Be Human

- **Creativity Unleashed:** With AI handling mundane tasks, individuals will have more time to pursue artistic expression, scientific inquiry, and innovation.
  - **Example:** AI-generated tools, such as OpenAI's DALL-E,

can assist artists in visualizing their ideas, while human creativity continues to guide the process.
- **Strengthening Connections:** AI-powered communication tools, such as real-time translators and virtual reality platforms, will help people maintain meaningful relationships across distances and cultures.

Time for Self-Discovery

In a post-work society, individuals will confront existential questions about what truly matters. AI's role in reducing stress and increasing leisure offers the chance to deepen personal exploration, spirituality, and mindfulness.

- **Example:** Universal Basic Income pilots, such as those conducted in Finland and Kenya, show that financial stability frees people to pursue personal passions, education, and volunteerism.

Collaboration as the Cornerstone of Progress

The challenges and opportunities of AI demand a collaborative, global response. Governments, businesses, educational institutions, and individuals must work together to ensure AI benefits humanity as a whole.

Global Cooperation for a Shared Future

- **International Standards for Ethical AI:** Organizations like UNESCO and the OECD are working to establish universal guidelines for fairness, transparency, and accountability in AI.
- **Addressing the Digital Divide:** Ensuring that AI tools and education reach underserved communities will be critical to creating an equitable future.

Empowering Individuals

- **Lifelong Learning:** Individuals must embrace continuous education to stay competitive and adaptable. Online platforms such as Coursera and Khan Academy are already democratizing access to AI education.
- **Advocacy and Participation:** Citizens can shape the future of AI by engaging in public debates, voting on policies, and supporting ethical initiatives.

Call to Action: Shape the Future

AI's future is not preordained—it is a creation of human choices. The time to act is now. Here's what each of us can do to ensure AI is a force for good:

1. **Educate Yourself:** Learn about AI's capabilities, risks, and implications to make informed decisions about its role in your life.
   - **Resource Example:** Take beginner-friendly courses like "AI for Everyone" by Andrew Ng on Coursera.
2. **Advocate for Ethical AI:** Support regulations and organizations that prioritize fairness, transparency, and inclusivity in AI development.
3. **Embrace Lifelong Learning:** Focus on building skills that complement AI, such as emotional intelligence, creativity, and adaptability.
4. **Engage in Dialogue:** Discuss AI with family, friends, and colleagues to raise awareness and build a shared vision for its future.

A Shared Vision for Prosperity

The age of AI is not about replacing humanity—it is about enhancing it. Together, we can build a world where AI amplifies human potential, deepens our connections, and unlocks opportunities for prosperity, creativity, and fulfillment. The path forward requires effort, courage, and collaboration, but the rewards—a more equitable, innovative, and enlightened world—are immeasurable.

As we stand on the brink of this transformative era, let us not fear the changes ahead but embrace them with purpose and vision. The future belongs to those who act today.

GLOSSARY OF TERMS

A

- **Algorithm**: A set of rules or instructions designed to solve a problem or perform a task. In AI, algorithms form the basis of machine learning models by processing data to produce outcomes.
- **Artificial General Intelligence (AGI)**: A hypothetical form of AI capable of performing any intellectual task that a human can do. Unlike narrow AI, AGI would possess general reasoning abilities, learning from experience across diverse tasks.
- **Artificial Intelligence (AI)**: The simulation of human intelligence by machines, enabling them to perform tasks like reasoning, learning, and decision-making.
- **Autonomous Systems**: Machines or technologies capable of performing tasks independently, without direct human intervention, often using AI for decision-making. Examples include self-driving cars and drones.

B

- **Bias (in AI)**: The presence of unfair or prejudiced outcomes in AI systems, often due to incomplete, unrepresentative, or biased training data. Bias can result in discriminatory decisions, especially in areas like hiring, policing, or loan approvals.
- **Big Data**: Extremely large datasets that are analyzed computationally to reveal patterns, trends, and associations. AI relies on big data to train machine learning models.

- **Black Box**: A term used to describe AI systems whose decision-making processes are not transparent or easily understood, making it difficult to interpret how outcomes are generated.

C

- **Chatbot**: A computer program designed to simulate human conversation through text or voice. Examples include customer service bots or conversational AI like ChatGPT.
- **Computer Vision**: A field of AI focused on enabling machines to interpret and analyze visual data, such as images and videos. Applications include facial recognition, object detection, and autonomous vehicles.
- **Consciousness (Machine)**: A speculative concept referring to whether machines could one day develop self-awareness or subjective experience. It is a topic of philosophical and ethical debate in AI development.
- **Creative AI**: AI systems designed to assist or independently produce creative outputs, such as music, art, writing, or design.
- **Cybersecurity**: The practice of protecting computers, networks, and data from digital attacks. AI is widely used in cybersecurity for detecting and preventing cyber threats.

D

- **Data Privacy**: The practice of safeguarding personal information from unauthorized access, misuse, or exploitation, especially by AI systems. This is a critical ethical concern in AI adoption.
- **Deep Learning**: A subset of machine learning that uses neural networks with multiple layers (hence "deep") to analyze and learn from large datasets. It powers advancements in image recognition, natural language processing, and autonomous vehicles.

- **Digital Divide**: The gap between those who have access to technology and digital resources and those who do not. Addressing this divide is crucial to ensuring equitable AI adoption.

E

- **Ethical AI**: AI systems that are designed and implemented in alignment with moral principles, prioritizing fairness, transparency, and accountability.
- **Explainable AI (XAI)**: AI systems that make their decision-making processes transparent and understandable to humans. This is critical for building trust and accountability in AI systems.

F

- **Facial Recognition**: A technology that uses AI to identify or verify a person's identity based on facial features. While it has practical applications, it raises ethical concerns about privacy and surveillance.
- **Fairness (in AI)**: The principle that AI systems should operate without bias or discrimination, treating all individuals equitably regardless of demographic factors.

G

- **Generative AI**: A type of AI that creates new content, such as text, images, or music, based on patterns in the data it has been trained on. Examples include ChatGPT and DALL-E.
- **Global AI Standards**: Internationally agreed-upon principles and regulations designed to ensure that AI is developed and used responsibly across the world.
- **General Data Protection Regulation (GDPR)**: A comprehensive privacy law in the European Union that governs how personal data is collected, processed, and

stored. It has implications for AI systems that handle personal data.

H

- **Human-in-the-Loop (HITL)**: An approach in AI system design where human oversight is integrated into the decision-making process, ensuring that critical choices align with human values and ethics.

I

- **Internet of Things (IoT)**: A network of interconnected devices that communicate with each other and collect data using sensors. AI is often used to analyze this data and automate processes.
- **Intelligent Automation**: The use of AI to automate processes and tasks, combining cognitive capabilities like learning and decision-making with robotic systems.

L

- **Lifelong Learning (AI)**: The concept that humans must continuously acquire new skills and knowledge to stay competitive in an AI-driven world.
- **Language Model**: A type of AI model trained to understand and generate human language. Examples include OpenAI's GPT models and Google's BERT.

M

- **Machine Learning (ML)**: A subset of AI where systems learn from data to improve their performance over time without being explicitly programmed.
- **Mass Surveillance**: The large-scale monitoring of individuals using AI-powered systems like facial recognition

and predictive analytics. It raises concerns about privacy and civil liberties.
- **Metaverse**: A collective virtual shared space, often enabled by augmented and virtual reality technologies, where people can interact with each other and digital environments. AI plays a critical role in creating immersive and personalized metaverse experiences.

N

- **Natural Language Processing (NLP)**: A field of AI focused on enabling machines to understand, interpret, and generate human language. Applications include chatbots, translation tools, and sentiment analysis.
- **Neural Network**: A type of machine learning model inspired by the structure of the human brain, consisting of layers of interconnected nodes that process data.

P

- **Predictive Policing**: The use of AI to analyze historical crime data and predict potential criminal activity. While it aims to improve resource allocation, it is often criticized for perpetuating systemic biases.
- **Privacy by Design**: A principle in system development that ensures privacy is considered and integrated from the outset of the design process.

R

- **Reinforcement Learning**: A machine learning technique where an AI agent learns to make decisions by interacting with an environment and receiving rewards or penalties for its actions.
- **Responsible AI**: The practice of designing and deploying AI systems that prioritize ethical principles, societal well-being,

and accountability.
- **Robotics**: A field of engineering and AI focused on designing and building machines capable of performing physical tasks, often autonomously.

S

- **Singularity (Technological)**: A hypothetical point in the future when AI surpasses human intelligence, leading to unprecedented technological and societal changes.
- **Supervised Learning**: A type of machine learning where a model is trained on labeled data, learning to map inputs to outputs based on examples.

T

- **Transparency (AI)**: The principle that AI systems should operate in ways that are understandable and explainable to users, ensuring accountability and trust.
- **Training Data**: The data used to train AI models, teaching them to recognize patterns, make predictions, or perform tasks.

U

- **Universal Basic Income (UBI)**: A proposed economic policy where all citizens receive a guaranteed income to offset the job displacement caused by automation and AI.
- **Unsupervised Learning**: A type of machine learning where a model identifies patterns and relationships in data without labeled outcomes.

W

- **Wearable AI**: Devices equipped with AI capabilities, such as fitness trackers or smartwatches, that monitor health metrics and assist in daily tasks.

X

- **Explainable AI (XAI)**: (See above) AI systems designed to make their processes transparent and understandable.

Z

- **Zero-Day Threat**: A cybersecurity term referring to a vulnerability in software or hardware that is exploited before the developer has had time to address it. AI plays a role in both identifying and preventing these threats.

www.ingramcontent.com/pod-product-compliance
Lightning Source LLC
Chambersburg PA
CBHW071037240526
45469CB00006BD/2236